Environmental Science and Engineering

Environmental Science

Series editors

Rod Allan, Burlington, Canada
Ulrich Förstner, Hamburg, Germany
Wim Salomons, Haren, The Netherlands

More information about this series at http://www.springer.com/series/3234

Riccardo Valentini · Franco Miglietta
Editors

The Greenhouse Gas Balance of Italy

An Insight on Managed and Natural Terrestrial Ecosystems

Springer

Editors
Riccardo Valentini
Department for Innovation in Biological, Agro-Food and Forest System (DIBAF)
University of Tuscia
Viterbo
Italy

and

Impacts on Agriculture, Forest and Natural Ecosystem Division (IAFENT)
Euro-Mediterranean Center on Climate Changes (CMCC)
Viterbo
Italy

Franco Miglietta
FoxLab Institute of Biometeorology
National Research Council of Italy (CNR) and Edmund Mach Foundation
San Michele all'Adige
Italy

ISSN 1431-6250
Environmental Science and Engineering
ISBN 978-3-642-32423-9 ISBN 978-3-642-32424-6 (eBook)
DOI 10.1007/978-3-642-32424-6

Library of Congress Control Number: 2014957323

Springer Heidelberg New York Dordrecht London

Printed on acid-free paper

Springer-Verlag GmbH Berlin Heidelberg is part of Springer Science+Business Media (www.springer.com)

Contents

Part I The Overview

1 **The Greenhouse Gas Balance of Italy: A Synthesis** 3
Maria Vincenza Chiriacò and Riccardo Valentini

2 **Carbon, Water and Energy Fluxes of Terrestrial Ecosystems in Italy** 11
Dario Papale, Mirco Migliavacca, Edoardo Cremonese, Alessandro Cescatti, Giorgio Alberti, Manuela Balzarolo, Luca Belelli Marchesini, Eleonora Canfora, Raffaele Casa, Pierpaolo Duce, Osvaldo Facini, Marta Galvagno, Lorenzo Genesio, Damiano Gianelle, Vincenzo Magliulo, Giorgio Matteucci, Leonardo Montagnani, Fabio Petrella, Andrea Pitacco, Guenther Seufert, Donatella Spano, Paolo Stefani, Francesco P. Vaccari and Riccardo Valentini

3 **Biogenic Volatile Organic Compound Emissions** 47
Rita Baraldi, Francesca Rapparini, Osvaldo Facini, Claudia Justina Kemper Pacheco, Giorgio Matteucci, Enzo Brancaleoni and Paolo Ciccioli

Part II Forests

4 **The Role of Managed Forest Ecosystem: An Inventory Approach** . . . 61
Anna Barbati and Piermaria Corona

5 **The Role of Managed Forest Ecosystems: A Modeling Based Approach** 71
Angelo Nolè, Alessio Collalti, Marco Borghetti, Marta Chiesi, Gherardo Chirici, Federico Magnani, Serena Marras, Fabio Maselli, Costantino Sirca, Donatella Spano and Riccardo Valentini

6 **Emissions from Forest Fires: Methods of Estimation and National Results** 87
Valentina Bacciu, Donatella Spano and Michele Salis

7 **Carbon Losses Due to Wood Harvesting and the Role of Wood Products** 103
Marco Marchetti, Gherardo Chirici and Bruno Lasserre

Part III Croplands, Grasslands and Natural Ecosystems

8 **Soil Carbon Stocks and Fluxes** 119
Tommaso Chiti, Costantino Sirca, Mirco Rodeghiero, Donatella Spano and Riccardo Valentini

9 **N_2O Emission Factors for Italian Crops** 135
Simona Castaldi, Giorgio Alberti, Teresa Bertolini, Annachiara Forte, Franco Miglietta, Riccardo Valentini and Angelo Fierro

10 **Cropland and Grassland Management** 145
Emanuele Lugato, Agata Novara, Damiano Gianelle, Loris Vescovo and Alessandro Peressotti

Part IV Regional Case Studies

11 **The Role of Vineyards in the Carbon Balance Throughout Italy** 159
Damiano Gianelle, Luciano Gristina, Andrea Pitacco, Donatella Spano, Tommaso La Mantia, Serena Marras, Franco Meggio, Agata Novara, Costantino Sirca and Matteo Sottocornola

12 **Afforestation and Reforestation: The Sicilian Case Study** 173
Juliane Rühl, Luciano Gristina, Tommaso La Mantia, Agata Novara and Salvatore Pasta

13 **Afforestation and Reforestation: The Friuli Venezia Giulia Case Study** 185
Giorgio Alberti, Gemini Delle Vedove, Silvia Stefanelli and Giuseppe Vanone

14 **Trying to Link Vegetation Units with Biomass Data: The Case Study of Italian Shrublands** 195
Salvatore Pasta, Tommaso La Mantia, Serena Marras, Costantino Sirca, Donatella Spano and Riccardo Valentini

Acronyms

AF/R/D	Afforestation/reforestation/deforestation
AIC	Akaike's information criterion
ANN	Artificial Neural Network
BEF	Biomass expansion factor
BVOC	Biogenic volatile organic compounds
BWD	Basic wood density
CAI	Current annual increment
CC	Combustion completeness
CE	Combustion efficiency
CLC	Corine land cover
CO_2 eq	Amount of CO_2 that would have the same global warming potential (GWP) for a given amount of greenhouse gas
CRO	Cropland
CUP	Carbon uptake period
DBF	Deciduous broadleaf forest
DBH	Diameter at breast height
EBF	Evergreen broadleaf forest
EF	Emission factor
ENF	Evergreen needleleaf forest
FE	Fire emissions
FIE	Fertilizer-induced emission
FM	Forest management
GHG	Greenhouse gas
GIS	Geographic information system
GPG-LULUCF	Good practice guidance for land use, land-use change and forestry. It refers to IPCC (2003) Penman J., Gytarsky M., Hiraishi T., Krug, T., Kruger D., Pipatti R., Buendia L., Miwa K., Ngara T., Tanabe K., Wagner F., Good Practice Guidance for Land Use, land-Use Change and Forestry IPCC/IGES, Hayama, Japan
GPP	Gross primary production

GRA	Grassland
IAV	Interannual variability
INFC	National inventory of forests and forest carbon pools
IPCC	Intergovernmental panel on climate change
ISCI	Carbon stock inventory
ISTAT	National statistics agency
IUTI	Land use inventory
LAI	Leaf area index
LUC	Land use change
LULUCF	Land use, land-use change and forestry
MBE	Mean bias error
NEE	Net ecosystem exchange
NEP	Net ecosystem production
NPP	Net primary production
Nr	Reactive nitrogen
PAR	Photosynthetically active radiation
PBMs	Process-based models
PFT	Plant functional types
QA/QC	Quality assurance/quality control
RMSE	Root mean square error
SHB	Shrubland
SLA	Specific leaf area
SOA	Secondary organic aerosols
SOC	Soil organic carbon
SOM	Soil organic matter
SVAT	Soil-vegetation-atmosphere transport
TER	Total ecosystem respiration
UNFCCC	United nations framework convention on climate change
VOC	Volatile organic compounds
VPD	Vapor pressure deficit

Part I
The Overview

Chapter 1
The Greenhouse Gas Balance of Italy: A Synthesis

Maria Vincenza Chiriacò and Riccardo Valentini

Abstract In this chapter a comprehensive assessment of the greenhouse gases budget of the Italian terrestrial ecosystems is provided, with particular attention to forest, cropland and grassland ecosystems and some case studies focusing on Italian shrublands and lands naturally or artificially converted to forests. Different methods have been applied and compared, such as regional measurements, use of flux networks and data-driven models within specific sectoral approaches in order to characterize the greenhouse gases budget of terrestrial ecosystems. The results presented respond also to the growing interest of the recent years in the role of the carbon cycle of terrestrial ecosystems and its relevance for national policies on mitigation and adaptation to climate changes.

1.1 Introduction

This chapter perceives the challenge to address in a comprehensive way the full greenhouse gases budget of the Italian terrestrial ecosystems, with particular attention to forest ecosystems, cropland and grassland ecosystems with some case studies focusing on Italian shrublands and lands naturally or artificially converted to forests.

The wealth of research information presented is mainly referred to the results of a national project, CarboItaly, which involved a number of Italian research institutions and several researchers with the aim to produce data and information useful to characterize different compartments of the greenhouse gases budget of the Italian terrestrial ecosystems, with a special emphasis on forest, croplands, grasslands and natural ecosystems.

M.V. Chiriacò (✉) · R. Valentini
Department for Innovation in Biological, Agro-Food and Forest System (DIBAF),
University of Tuscia, Via San C. de Lellis s.n.c., 01100 Viterbo, Italy
e-mail: chiriaco@unitus.it

M.V. Chiriacò · R. Valentini
Impacts on Agriculture, Forest and Natural Ecosystem Division (IAFENT),
Euro-Mediterranean Center on Climate Changes (CMCC), Viterbo, Italy

R. Valentini and F. Miglietta (eds.), *The Greenhouse Gas Balance of Italy*,
Environmental Science and Engineering, DOI 10.1007/978-3-642-32424-6_1

The results provided in this book respond to the growing interest of the recent years in the role of the carbon cycle of terrestrial ecosystems and its relevance for national policies on mitigation and adaptation to climate changes.

There is also a growing need on the part of institutions, agencies and policy stakeholders for new data and analysis in relation to the United Nations Framework Convention on Climate Change (UNFCCC) process. In particular, the data presented in this book contribute to build a basis for a full carbon accountability of the land sector.

1.2 Methods for the assessments of the greenhouse gases budget

Different methods have been applied and compared, such as regional measurements, use of flux networks and data-driven models within specific sectoral approaches in order to characterize the greenhouse gases budget of terrestrial ecosystems. The Eddy Covariance technique has been applied to a range of Italian sites representative of the main plant functional types (PFT) and of the main Italian macro-regions, to provide data and measures of carbon dioxide, water and energy flow for the whole period of the project. This has made Italy the country with the most dense network of production and eddy covariance Europe. The CarboItaly sites also contributed to the activities in the global network FLUXNET (www.fluxdata.org). Measures of the NEE and its partitioning into the Gross Primary Production (GPP) and ecosystem respiration have been performed with the Eddy Covariance technique in order to analyse the role of the Carbon Uptake Period in the total Net Ecosystem Exchange (NEE) and define the effect of temperature and precipitation, disturbances and management practices on the interannual variability of NEE, aiming at a better understanding of the role and response of ecosystems to climate change.

The role of the Italian managed forest ecosystems in national greenhouse gases budget has been widely discussed and investigated in this book. Integration of land use and forest inventory approaches as well as modeling based approaches have been applied for the assessment of the national forest Gross and net primary productivity (GPP and NPP) and the carbon, water, and elemental cycles. In forest ecosystems firewood and forest harvesting represent a net carbon loss but the use of wood, carbon-neutral renewable resource, for generating energy has also a strong substitution effect as it avoids the use of fossil fuels which are highly CO_2 emitting. The use of wood for construction purposes, substituting traditional materials, tends to increase carbon sequestration and to contribute to climate change mitigation. The great potential of the increasing use of forest wood products for energy, building and furniture purposes in contributing to the reduction of GHG emissions and to a more sustainable development have also been investigated. Another important aspect in the terrestrial carbon budget is represented by the emissions from forest fires. Several experimental and modelling studies have been conducted and presented in this book to improve knowledge of the atmospheric impact of vegetation fires. Biogenic Volatile Organic Compounds (BVOCs) fluxes have been measured in forest CarboItaly sites and a GIS-based model has been developed for predicting BVOCs emissions from the Italian forest ecosystems in order to estimate the fraction of the Net Ecosystem Production lost as reduced carbon and to assess the impact of BVOCs in the formation of ozone and secondary organic aerosols.

The cropland ecosystems with the organic carbon stock in mineral soils and in permanent woody structure of perennial tree crops also play a key role in the terrestrial carbon budget. In this book an assessment of the total soil organic carbon (SOC) stock in the top 30 cm of mineral soil for the whole Italian territory, according to the different land use types of the Intergovernmental Panel on Climate Change (IPCC) cropland category (arable land, agroforestry, vineyards, olive groves, orchards and rice fields), has been performed as a basis for future land use scenarios and to address mitigation policy at country level.

Also nitrous oxide (N_2O) emissions from several Italian croplands along a latitudinal gradient as well as the relationship between N_2O production and applied N fertilization rate were analyzed and empirically derived from experimental sites in order to assess the effect of climate variability on N_2O emissions for Italian crops and to verify the reliability of internationally applied emission factors for temperate crops.

The effects of climate and management practices on the carbon balance in the soil and in the perennial woody biomass have been particularly investigated for Italian vineyard ecosystems. Moreover, regional case studies have been focused of different managements in cropland and grassland ecosystems and on the consequence on the effect on the soil organic carbon of the abandonment of agricultural land and the natural or artificial afforestation.

1.3 Results

The main results presented in the next chapters have been grouped and further elaborated in order to provide a comprehensive assessment of the whole national greenhouse gases budget of terrestrial ecosystems. The values of Gross Primary Production (GPP) and Net Ecosystems Production (NEP) referred to the total Italian surface as well as the main national emissions of BVOCs and N_2O and CH_4 have been assessed and are reported in the Tables 1.1, 1.2 and 1.3.

Table 1.1 Carbon fixed in Italian natural ecosystems

Parameter	Unit value (g C m^2 y^{-1})	Total carbon uptake from natural ecosystems (Mt C y^{-1})	Comments
NEE	−227 (25 %: −264; 75 %: −191)	62.3	Median and interquartile range of empirical upscaling using artificial neural network (ANN) trained with eddy covariance data (Luyssaert et al. 2012) (Chap. 2). The national area of natural ecosystems includes forestland, cropland and grassland, and has been derived from Corine Land Cover 2006 (CLC 2006)
GPP	849 (25 %: 811; 75 %: 891)	233	

Table 1.2 Total Net Primary Production (NPP) of Italian forest ecosystems

Parameter	Unit value (g C m^2 y^{-1})	Total carbon fixed in Italian forest ecosystems (Mt C y^{-1})	Comments
NPP (only aboveground biomass, 2005)	135–145	11.82–12.70	Inventory approach (Chap. 4) Italian forest area from INFC (2005)
NPP (only below ground biomass, 2005)	40.5435	3.55–3.81	Based on inventory approach (Chap. 4) and elaborated considering a mean value of the root/shoot ratio (R) of the main Italian forest types equal to 0.30 (Federici et al. 2008) Italian forest area from INFC (2005)
NPP forest tree biomass (below + above ground) in 2005	176–189	15.37–16.51	
Simulated NPP (all forest carbon pools) total	300–450	26.72–40.08	Modeling approach (Chap. 5) The national area of Italian forest types has been derived from Corine Land Cover 2006 (CLC 2006)
• Mediterranean shrub land	317	6.67	
• Holm oak and evergreen woods	420	3.02	
• Woods mainly planted with Mediterranean pine trees and/or cypresses	368	0.73	
• Hygrophilous forests	320	0.23	
• Broad-leaved woods and plantations with non native species	302	3.17	
• Deciduous mixed oaks woods	402	7.98	
• Chestnut woods	467	3.54	
• Beech forests	381	3.5	
• Woods mainly planted with pine-trees in the sub-alpine and alpine areas (Silver fir and red fir woods)	407	3.52	
• Black pine and mountain pine woods	372	0.81	
• Conifers woods and plantations of non native species	392	0.09	

Table 1.3 Soil organic carbon (SOC) stock in Italian cropland ecosystems

Parameter	Unit value (g C m^2)	Total SOC stock in Italian croplands (Mt C)	Comments
SOC stock in the upper 30 cm	–	489 ± 148.2	Data referred to 2007 (Chap. 8)
SOC stock in the upper 30 cm	–	470.5	Projected for 2020, with no mitigation options (Chap. 8)
Arable land and Agroforestry	4.010 ± 230 to 6.980 ± 2530	285–497	SOC stock in Italian cropland sub-categories, depending on climate types (Chiti et al. 2010) Data referred to 2000 (Chap. 8)
Vineyards	3.920 ± 1.000 to 8.220 ± 5.490	281–590	
Olive groves	4.210 ± 1.330 5.600 ± 3.500	45.5–60.5	
Orchards	3.810 ± 990 to 5.780 ± 1.640	240–365	
Rice fields	6.010 ± 1.110 to 23.440 ± 8.030	12.8–19.9	

Table 1.4 Carbon stock and fluxes measured in local case studies in Italy

Parameter	Total value	Comments
Annual NEE Vineyards	89–145–814	Measured on sites of Serdiana (CA), Valle dell'Adige (TN) and Negrisia (TV) respectively, in 2009 NEE expressed in g C m^2 y^{-1} It includes all carbon pools: soil, grass cover, vine canopy and woody biomass (Chap. 11)
SOC stock permanent irrigated and irrigated arable lands	3.160	SOC Stock expressed in g C m^2 Measured in Sardinia croplands in 2008 (Chap. 8)
SOC stock vineyard	3.380	
SOC stock olive groves	3.220	
Aboveground carbon stock on afforested croplands	2530–4.030	SOC stock measured in g C m^2 on afforested cropland due to land use change after abandonment in Friuli Venezia Giulia (Chap. 13)
Belowground carbon stock on afforested croplands	660–1.050	
SOC stock on afforested croplands	5.430–7.700	
Aboveground NPP on afforested croplands	240–940	
Belowground NPP on afforested croplands	60–240	
Soil carbon sequestration on afforested croplands	40–110	

The total carbon fixed in the Italian natural ecosystems through the process of photosynthesis, expressed as GPP, can be assumed of 233 millions of tons of Carbon (Mt C), of which 62.3 Mt C represents the total Italian net carbon sink, expressed as NEE (Table 1.1).

A considerable amount of national carbon budget of natural ecosystems is represented by the Net Primary Production (NPP) of Italian forests, with a total of 15.37–16.51 millions of tons of Carbon per year (Mt C y^{-1}) fixed in the above and below ground biomass of forest trees and 26.72 (±5.34)–40.08 (±8.02) Mt C y^{-1}) if considering all carbon pools including litter, dead wood and forest soils (Table 1.2).

Another considerable amount of carbon contained in natural ecosystems is represented by the soil carbon (SOC) stock of Italian croplands. The SOC stock in the upper 30 cm of Italian agricultural soils has been assessed of about 489 (±148.2) Mt C in 2007 and is projected for 2020 = 470.5 Mt C, with no mitigation options (Table 1.3).

Others values of carbon stock and fluxes are reported in the Table 1.4 related to specific local case studies of the Italian territory.

As reported in Table 1.5, a fraction of the total Net Ecosystem Production (NEP) of the Italian forest ecosystems is lost as reduced carbon and transformed to BVOCs emissions assessed for 2006 = 41.2 Gg y^{-1} of Monoterpene emissions and to 31.7 Gg y^{-1} of Isoprene emissions. The BVOCs emissions from the Italian forest ecosystems in 2006. The loss of NEP from the Italian forest ecosystems as Biogenic Volatile Organic Compound emissions is estimated of about 3–4 %.

Moreover, another relevant annual loss of NEP of the Italian forest ecosystems is represented by the carbon emissions due to wood removals (Table 1.6).

The two most important non CO_2 greenhouse gases (GHG) exchanged with the atmosphere in Italian agricultural soils are the nitrous oxides (N_2O) and methane (CH_4).

Table 1.5 Total Italian Biogenic Volatile Organic Compounds (BVOCs) emissions from the Italian forest ecosystems in 2006

Parameters	Total emissions of BVOCs (Gg y^{-1})	Comments
Monoterpene emissions	41.2	of which: α-pinene (9.10 Gg y^{-1}), sabinene (4.34 Gg y^{-1}) and β-pinene (3.37 Gg y^{-1}) Limonene, myrcene, *trans*- and *cis*-beta ocimene, linalool, 1-8 cineol, camphene, β-phellandrene and terpinolene contributed for the rest of the emission (Chap. 3)
Isoprene emissions	31.7	

Table 1.6 Annual loss of NEP as carbon emissions due to wood removal from Italian forests

Parameters	Total emissions (Mt CO_2)	Comments
Annual carbon loss due to wood removal in Italy	6	Statistical data approach (Chap. 7)
Annual emission of hwp	0.92	Production approach (Chap. 7)

An emission of N_2O of 1.52 (± 0.04) Mt CO_2 eq yr^{-1} and a slight sink of CH_4 (−0.08 ± 0.001 Mt CO_2 eq yr^{-1}) has been assessed for the five main crops which represented 54% of the total harvested land in 2009, excluding rice paddies (Lugato et al., 2010). Applying the emission factor (EF, kg N_2O-N/kg extra N) of 0.8% (derived from the range 0.4 to 0.8% provided by Chapter 9 and 10) to the total amount of fertilizer N consumed in Italy in 2009 (514.480 tons of N) (FAOSTAT 2014) and considering a similar flux strength for the remaining Italian harvested land, the total amount of N_2O emissions in CO_2 equivalent is 1.93 (± 0.09) Mt CO_2 eq yr^{-1} for 2009 while the total sink of CH_4 (excluding rice cultivation) is −0.148 (± 0.002) Mt CO_2 eq yr^{-1} (Table 1.7).

Table 1.8 shows SOC stock variations related to different management practices, ranging from a carbon sequestration potential of 26–67 g C m^2 y^{-1} when

Table 1.7 Fluxes of N_2O and CH_4, expressed as CO_2 equivalents, in soils of Italian croplands and woodland crops for year 2009 (rice paddies excluded)

Parameters	Total fluxes of GHG (Mt CO_2 eq)	Comments
N_2O	1.93 (± 0.09)	Estimated applying an EF of 0.8 % to total N fertilizer consumption in 2009 (Chapter 9 and 10)
CH_4	-0.148 (± 0.002)	Sum of −0.08 Mt CO_2 eq from modeling simulation applied to 54 % of national harvested land, and −0.068 Mt CO_2 eq extrapolated applying a similar GHG source strength (tons CO_2 eq/ha) to the remaining harvested land (excluding ricepaddies) (Lugato et al 2010).

Table 1.8 SOC stock variations related to different management practices (the negative sign indicates a sequestration potential)

Parameters	Total carbon (g C m^2 y^{-1})	Comments
SOC—Average annual carbon loss (with no mitigation options)	20–50	SOC loss in Italian croplands Gardi and Sconosciuto (2007), Janssens et al. (2005), Lugato et al. (2010), Morari et al. (2006) (Chap. 8)
SOC—Annual carbon loss (with no mitigation options)	16	Calculated as difference between 1990 and 2000 (Chap. 8)
Permanent set a side or zero tillage	−40	Carbon sequestration potential in soils of mitigation options in Italian croplands. Freibauer et al. (2004), Smith et al. (2000a, b) (Chap. 8)
Perennial crops or deep rooting crops	−60	
Change from conventional to organic farming	−50	
Best management practice: change from cropland to grassland; no-till; farm yard manure	−26 to −67	Field experiments (Morari et al. 2006; Lugato et al. 2006; Triberti et al. 2008; Mazzoncini et al. 2011) (Chap. 10)
Abandonment of a vineyard	+27 %	SOC stock variation in Sicilian vineyards (Chap. 11)
Re-planting	−43 % initially	

best management practices like permanent set a side or zero tillage, perennial crops or deep rooting crops, change from conventional to organic farming change and from cropland to grassland, no-till and use of farm yard manure are applied.

References

Chiti T, Papale D, Smith P, Dalmonech D, Matteucci G, Yeluripati J, Rodeghiero M, Valentini R (2010) Predicting changes in soil organic carbon in mediterranean and alpine forests during the Kyoto Protocol commitment periods using the century model. Soil Use Manag 26:475–484. doi: 10.1111/j.1475-2743.2010.00300.x, ISSN: 0266-0032

CLC (2006) Corine Land Cover 2006 technical guidelines: European environment agency, EEA technical report No 17/2007, p 66

FAOSTAT (2014) Consumption of fertilizer N in Italy. http://faostat3.fao.org/home/index.html

Federici S, Vitullo M, Tulipano S, De Lauretis R, Seufert G (2008) An approach to estimate carbon stocks change in forest carbon pools under the UNFCCC: the Italian case. IForest 1:86–95

Freibauer A, Rounsevell MDA, Smith P, Verhagen J (2004) Carbon sequestration in the agricultural soils of Europe. Geoderma 122:1–23

Gardi C, Sconosciuto F (2007) Evaluation of carbon stock variation in Northern Italian soils over the last 70 years. Sustain Sci 2:237–243

INFC (2005) Inventario nazionale delle foreste e dei serbatoi forestali di carbonio: estensione e composizione dei boschi (http://www.sian.it/inventarioforestale/jsp/dati_introa.jsp)

Janssens IA, Freibauer A, Schlamadinger B, Ceulemans R, Ciais P, Dolman AJ, Heimann M, Nabuurs GJ, Smith P, Valentini R, Schulze ED (2005) The carbon budget of terrestrial ecosystems at country-scale-a European case study. Biogeosciences 2:15–26

Lugato E, Berti A, Giardini L (2006) Soil organic carbon (SOC) dynamics with and without residue incorporation in relation to different nitrogen fertilisation rates. Geoderma 135:315–321. doi:10.1016/j.geoderma.2006.01.012

Lugato E, Zuliani M, Alberti G, Delle Vedove G, Gioli B, Miglietta F, Peressotti A (2010) Application of DNDC biogeochemistry model to estimate greenhouse gas emission from Italian agricultural areas at high spatial resolution. Agr Ecosyst Environ 139:546–556

Luyssaert S, Abril G, Andres R, Bastviken D, Bellassen V, Bergamaschi P, Bousquet P, Chevallier F, Ciais P, Corazza M, Dechow R, Erb K-H, Etiope G, Fortems-Cheiney A, Grassi G, Hartmann J, Jung M, Lathiere J, Lohila A, Mayorga E, Moosdorf N, Njakou DS, Otto J, Papale D, Peters W, Peylin P, Raymond P, Odenbeck CR, Saarnio S, Schulze E-D, Szopa S, Thompson R, Verkerk PJ, Vuichard N, Wang R, Wattenbach M, Zaehle S (2012) The European land and inland water CO_2, CO, CH_4 and N_2O balance between 2001 and 2005. Biogeosciences 9(8):3357–3380. doi: 10.5194/bg-9-3357-2012, ISSN: 1726-4170

Mazzoncini M, Bahadur Sapkota T, Barberi P, Antichi D, Risaliti R (2011) Long-term effect of tillage, nitrogen fertilization and cover crops on soil organic carbon and total nitrogen content. Soil Tillage Res 114:165–174. doi:10.1016/j.still.2011.05.001

Morari F, Lugato E, Berti A, Giardini L (2006) Long term effects of recommended management practices on soil carbon changes and sequestration in north-eastern Italy. Soil Use Manag 22:71–81

Smith P, Powlson DS, Smith JU, Falloon P, Coleman K (2000a) Meeting Europe's climate change commitments: quantitative estimates of the potential for carbon mitigation by agriculture. Glob Change Biol 6:525–539

Smith WN, Desjardins RL, Patty E (2000b) The net flux of carbon from agricultural soils in Canada 1970–2010. Glob Change Biol 6:557–568

Triberti L, Nastri A, Giordani G, Comellini F, Baldoni G, Toderi G (2008) Can mineral and organic fertilization help sequestrate carbon dioxide in cropland? Eur J Agron 29:13–20

Chapter 2
Carbon, Water and Energy Fluxes of Terrestrial Ecosystems in Italy

Dario Papale, Mirco Migliavacca, Edoardo Cremonese, Alessandro Cescatti, Giorgio Alberti, Manuela Balzarolo, Luca Belelli Marchesini, Eleonora Canfora, Raffaele Casa, Pierpaolo Duce, Osvaldo Facini, Marta Galvagno, Lorenzo Genesio, Damiano Gianelle, Vincenzo Magliulo, Giorgio Matteucci, Leonardo Montagnani, Fabio Petrella, Andrea Pitacco, Guenther Seufert, Donatella Spano, Paolo Stefani, Francesco P. Vaccari and Riccardo Valentini

Abstract In this chapter the Eddy Covariance network of Italy is presented, with a short introduction to each of the 29 sites that were active during the CarboItaly project. These sites provided a unique dataset for a better study and understanding of the carbon cycle of terrestrial ecosystems and the links between carbon sink capacity and the main environmental factors. After a number of examples of Eddy Covariance time series where it is possible to see the effect of interannual climate variability and disturbances and managements practices, an analysis of the role of

D. Papale (✉) · M. Balzarolo · L. Belelli Marchesini · E. Canfora · P. Stefani · R. Valentini
Department for Innovation in Biological, Agro-Food and Forest System (DIBAF), University of Tuscia, Viterbo, Italy
e-mail: darpap@unitus.it

M. Migliavacca · A. Cescatti · G. Seufert
Joint Research Centre, Institute for Environment and Sustainability, European Commission, Ispra, Italy

E. Cremonese · M. Galvagno
ARPA Valle D'Aosta, Aosta, Italy

G. Alberti
Department of Agricultural and Environmental Sciences, University of Udine, Udine, Italy

R. Casa
DAFNE, University of Tuscia, Viterbo, Italy

O. Facini
IBIMET, CNR, Bologna, Italy

L. Genesio · F.P. Vaccari
IBIMET, CNR, Florence, Italy

D. Gianelle
Department of Sustainable Agro-Ecosystems and Bioresources, Research and Innovation Centre, Fondazione Edmund Mach, Michele All' Adige, Trento, Italy

R. Valentini and F. Miglietta (eds.), *The Greenhouse Gas Balance of Italy*, Environmental Science and Engineering, DOI 10.1007/978-3-642-32424-6_2

the Carbon Uptake Period in the total Net Ecosystem Exchange (NEE) definition and a study of the effect of temperature and precipitation on the interannual variability of NEE are presented in order to show the way these data can contribute to a better understanding of the role and response of ecosystems to climate change.

2.1 Introduction

Monitoring carbon, water and energy fluxes between terrestrial ecosystems and atmosphere is essential for a better understanding of the biological and ecological processes, also in relation to climate variability and climate change, and to assess the carbon balance of the different ecosystems and their ability to sequester CO_2 from the atmosphere.

In order to improve the understanding of the quantities involved in the carbon balance of terrestrial ecosystems, it is important to provide the following definitions: it is defined Gross Primary Production (GPP) of an ecosystem, the total amount of CO_2 that is fixed by the vegetation in photosynthesis. The synthesis of new plant tissues and the maintenance of the plants themselves require energy that is provided by the autotrophic respiration (Ra). The difference between the amount of carbon fixed by photosynthesis and respired by the vegetation is defined Net Primary Production (NPP):

$$NPP = GPP - Ra$$

V. Magliulo
ISAFOM, CNR, Ercolano, NA, Italy

G. Matteucci
ISAFOM, CNR, Rende, CS, Italy

G. Matteucci
IBAF, CNR, Monterotondo, RM, Italy

L. Montagnani
Forest Services and Agency for the Environment, Autonomous Province of Bolzano,
Faculty of Science and Technology, Free University of Bolzano, Bolzano, Italy

F. Petrella
Institute for Wood Plant and Environment (IPLA),
Environmental Disease and Soil Conservation Unit, Turin, Italy

A. Pitacco
DAFNAE, University of Padua, Padua, Italy

D. Spano
Department of Science for Nature and Environmental Resources (DIPNET),
University of Sassari, via E. de Nicola 9, 07100 Sassari, Italy

P. Duce
IBIMET, CNR, Sassari, Italy

D. Spano
Impacts on Agriculture, Forest and Natural Ecosystem Division (IAFENT),
Euro-Mediterranean Center on Climate Changes (CMCC), Sassari, Italy

The NPP is allocated to the production of biomass (wood, leaves, roots, fruits, seeds etc.) and respired back to the atmosphere mainly due to decomposition by microbial activities. The quantity of carbon lost by respiration for heterotrophic organisms is defined heterotrophic respiration (Rh) and the difference between NPP and Rh is the Net Ecosystem Production

$$NEP = NPP - Rh = GPP - TER$$

The NEP represents the net ecosystem carbon sink or source due to physiological processes and it is also named Net Ecosystem Exchange (NEE) when it is quantified using measurements of CO_2 exchanges between ecosystem and atmosphere (while it is called NEP when measured using inventory approaches). The two terms are somehow interchangeable but in general with opposite signs: a flux of carbon from the atmosphere to the ecosystem is positive in NEP and negative in NEE. The sum of the two respiration components (Rh and Ra) represents the total ecosystem respiration (TER).

The NEP is however different from the long term carbon balance of the ecosystem because there can be changes in the carbon stocks due to episodic losses by natural or anthropogenic disturbances and management practices. For this reason the Net Biome Production (NBP) is defined as (Schulze et al. 2000):

$$NEP = NPP - CO_2 \text{ losses due to disturbances}$$

There are different possible approaches and methods to measure the fluxes of energy and greenhouse gases (GHGs) in terrestrial ecosystems, ranging from inventory approaches to chambers measurements and ecosystem scale techniques such as the Eddy Covariance method (Aubinet et al. 2012).

The Eddy Covariance methodology has been developed in the early '90s and has been widely applied at global level. It is based on high frequency (10 Hz) measurements of wind speed, temperature and gas concentration using a three-axis sonic anemometer (which measures the wind speed along the three axis) and a fast response gas analyzer, typically an Infra Red Gas Analyzer (IRGA) for CO_2 and H_2O, even if new systems have been recently developed and commercialized to measure high frequency concentrations of other gases such CH_4, N_2O and O_3.

With the Eddy Covariance technique it is possible to measure the Net Ecosystem Exchange (NEE) of a GHG of a given surface extended around the monitoring tower (the footprint). The extension and shape of the footprint is function of the wind speed, wind direction and the difference between the measurement and canopy heights and it has generally a radius between few hundred meters and one kilometer around the measurement point. The Eddy Covariance technique is the only method available today to continuously measure the net ecosystem exchanges at ecosystem level and in a not-destructive way.

In addition, for CO_2 NEE measurements, there is the possibility to statistically partition the net carbon fluxes measured into its major components as the gross primary production (GPP) and the ecosystem respiration (Reichstein et al. 2005; Lasslop et al. 2010) allowing a better interpretation of the fluxes in terms of ecosystem processes.

Nowadays, more than 500 sites exist globally, organized in regional networks contributing to the global network FLUXNET (http://fluxnet.ornl.gov/) with the aim to create global standardized datasets of Eddy Covariance measurements available to the scientific community (Papale et al. 2012). The usefulness of these measurements has been proved by the large range of applications published in the last years, ranging from empirical up scaling (Jung et al. 2010; Beer et al. 2010) to climate-ecosystem interactions (Reichstein et al. 2007), ecosystem functioning (Mahecha et al. 2010; Williams et al. 2012) and model and remote sensing products parameterization and validation (Maselli et al. 2009; Chiti et al. 2010; Migliavacca et al. 2011; Wang et al. 2012).

2.2 The Italian Network

In Italy the first sites measuring CO_2 and H_2O fluxes continuously using the Eddy Covariance technique were started in 1996–1997 (IT-Cpz, IT-Col and IT-Ren) within the context of the EUROFLUX European project (ENVCT 0095-0078), but the number increased rapidly in the following years thanks to other European research projects supporting the continental network of sites.

The CarboItaly FISR Italian project offered the opportunity to consolidate, enlarge and standardize the Italian network of sites which amounted to 29 sites in the 2007–2010 period. These sites, representing different Plant Functional Types (PFTs) and distributed along the Italian peninsula (Fig. 2.1), are briefly described in the following subsections. The network has been coordinated by the University of Tuscia, also hosting the database where all the data have been processed, stored and are today available to its users.

The data measured with the Eddy Covariance technique need a multiple-steps processing that, in particular for CO_2 and H_2O, has been standardized and consolidated also during the CarboItaly project. The developed processing chain is applicable to all those sites minimizing potential differences due to the data Quality assurance/Quality control (QA/QC) used in particular when synthesis activities involving multiple sites are performed. The measurements acquired by the Italian sites have all been centrally processed according to the international standards described in Papale et al. (2006), Moffat et al. (2007) and Reichstein et al. (2005).

2.2.1 Amplero (IT-Amp)

The Amplero Mediterranean mountainous grassland is located in central Italy in the Abruzzo region (41.90409° N; 13.60516° E) on a flat to gently south sloping (2–3 %) area, at 884 m.a.s.l. The experimental area is a homogenous mixed grassland, mainly composed by few dominant graminoids (genus *Poa* 10 %), forbs (genus *Trifolium* 30 %, genus *Medicago* 20 %) and composites (genus *Geranium* 20 %, genus *Cerastium* 20 %). The growing season ranges from April to the end of May (roughly 60 days long) with the maximum of production at the end of May, while the

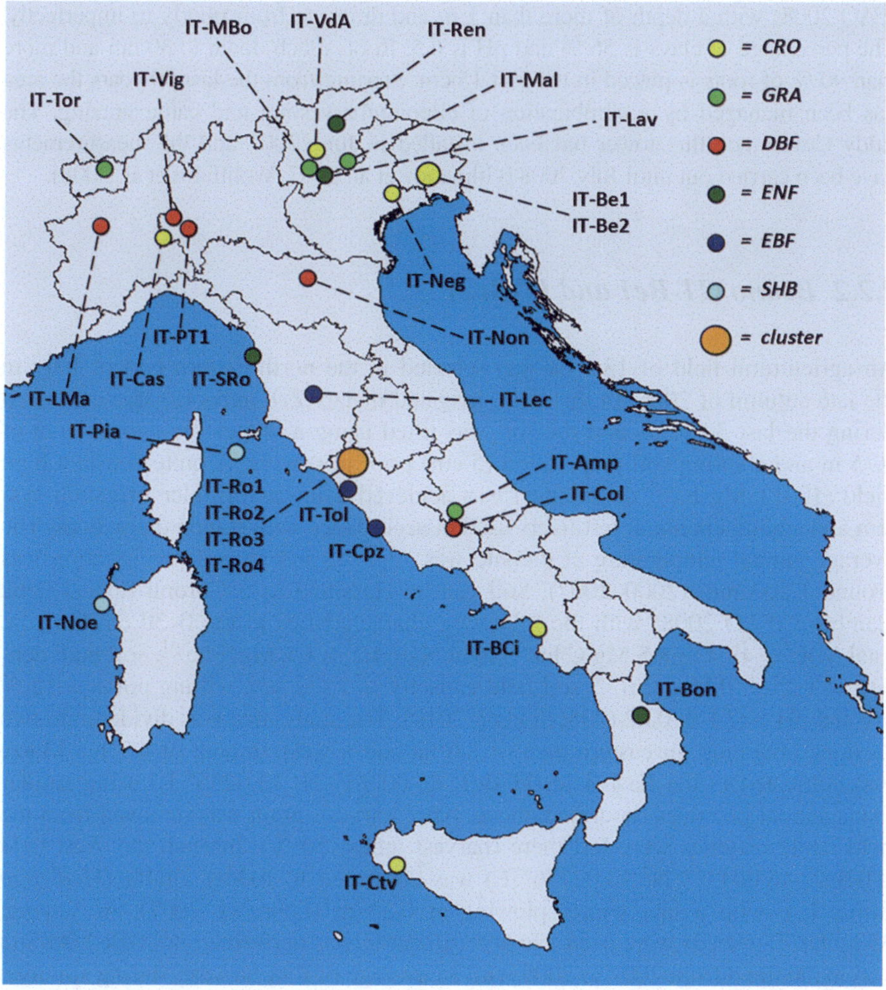

Fig. 2.1 Map of the Italian Eddy Covariance sites contributing to the CarboItaly project. *Colors* represent the different PFTs: *CRO* cropland, *GRA* grassland, *DBF* deciduous broadleaf forest, *ENF* evergreen needleleaf forest, *EBF* evergreen broadleaf forest, *SHB* shrubland. The Roccarespampani cluster includes two *DBF* and two *CRO* sites

senescence occurs in June. At the end of June there is a clip and after that the animal grazing (cows, horses and donkeys) until the end of November. The estimated stocking rate is around 0.5–1 animals per hectare each year and we can therefore assume that the grazing is extensively managed. The climate of this area is mountainous Mediterranean (Petriccione et al. 1993), typical climate of the Italian Apennines region. Average annual precipitation is 1,365 mm (ARSSA local database) with two precipitation peaks: one in autumn (October to December) and the other one in spring (March to April). The precipitation minimum occurs in July. The soil is Haplic Phaeozems

(FAO 2008) with a depth of more than 1 m and drainage from poorly to imperfectly. The percentage of clays is 56 % and pH is 6.5. Roots reach down to 30 cm and more than 90 % of roots is placed in the first 15 cm. Starting from the last 50 years the area has been managed by a combination of clipping/harvesting and cattle grazing. The Eddy Covariance flux tower has been installed in June 2002 and the measurements have been carried out until July 2008 (Gilmanov et al. 2007; Wohlfahrt et al. 2008).

2.2.2 Beano (IT-Be1 and IT-Be2)

An agricultural field of 13.3 ha was selected in the north-eastern part of Italy in the late autumn of 2006. In this field, irrigated maize (*Zea mays* L.) was cultivated during the last 30 years and the soil was tilled using a winter plow to a depth of 0.35 m and a spring soil preparation (5 cm) prior to sowing. A quite constant high yield (10–11 Mg ha^{-1} dry matter) was achieved using a sprinkler irrigation system and adding chemical fertilizers used in accordance with standard practices. The average annual temperature at the site was 13.7 °C and annual precipitation was around 1,200 mm (2000–2007). Soil can be classified as a Chromi-Endoskeletic Cambisol (FAO 2008) with the following characteristics in the 0–30 cm horizon: total SOC = 48.4 ± 8.5 Mg C ha^{-1}, total N = 4.2 ± 1.1 MgN ha^{-1}, soil bulk density = 1.25 ± 0.15 g cm^{-3}, soil field capacity = 23 % v/v, wilting point = 12 % v/v and pH = 7.1 ± 0.02 (Alberti et al. 2010). The study area was divided into two sections obtaining an eastern area of 8.6 ha and a western area of 4.7 ha. Maize was cultivated in the East field (IT-Be1, 46.00361° N; 13.0256° E) using similar management practices described above. After harvest, grain was removed from the field while residues were left there (harvest index: 50 %). Instead, the West field (IT-Be2, 46.00431° N; 13.02776° E) was converted to fodder alfalfa (*Medicago sativa* L.), with a late winter plowing at 0.35 m (February 2007). At sowing, 39 kg ha^{-1} of seeds were used. No N fertilization was performed in IT-Be2. Alfalfa was maintained from 2007 to 2009 with an average of four harvests during summer. In 2010, IT-Be2 was converted again to maize with a minimum tillage (0–5 cm) before sowing and following the same management practices used in IT-Be1.

2.2.3 Bonis (IT-Bon)

The experimental site is located in Southern Italy, in the Sila Greca mountain range, in the region of Calabria (39.47778° N; 16.53472° E). The watershed area is 140 ha, ranging from 975 to 1,300 m.a.s.l. and has been reforested between 40 and 50 years ago with conifers (*Pinus nigra* J.F. Arnold subsp. *laricio*, 80 %) and broad-leaves (*Castanea sativa* L., 6 %). The remaining surface is covered by pastures, crops, creeks and bare soil. The area is characterized by a Mountain Mediterranean climate with variable but generally sufficient rainfall (1,170 mm) and an average temperature of 8.7 °C. Since 1986, the watershed has been equipped with meteorological stations and instrumentation in order to monitor hydrometrical heights and river discharge. Components of the hydrological balance are studied since 1994 in forest structures characterized by different thinning regimes. Since 1999, the effect of different fire types on soil erosion and hydrology is also studied. The flux stations have been established in 2003 over 35 years-old pine plantation (1,175 m.a.s.l.) with 637 trees ha^{-1}. Mean diameter and height are 29.3 cm and 20 m, respectively, with a basal area of 45.4 m^2 ha^{-1}. In 2006, all sided Leaf Area Index (LAI) was 6.55 $\pm$ 0.18. Substrate is granite and soils are Ultic Haploxeralfs.

2.2.4 Borgo Cioffi (IT-BCi)

The Borgo Cioffi monitoring site has been running since July 2002 on a arable field located on Gaetano Iemma's farm (Eboli, Salerno, 40.52375° N; 14.95744° E), covering a total area of 150 ha in the centre of the Sele river Plain, an alluvial plain featuring an intensive and high income agricultural activity. The station is located about 20 m.a.s.l., which can be found south-west at a distance of about 5 km. The 16 ha field, in which all the experimental activities are carried out, is surrounded by irrigated land, has no aerodynamic obstacles in the vicinity of the boundaries able to influence the vertical profiles of atmospheric properties and it is irrigated by means of an automatic central pivot system.

The farm hosts a water buffalo operation with over 800 adult cattle, and the fields have been cultivated for decades to forage crops (alfalfa, winter and spring grass, corn silage) and winter vegetables (fennel, cauliflower). Solid and liquid manure are frequently applied at high rates.

The station is located approximately in the centre of the field, featuring an approximately rectangular shape with sides of 270 and 600 m. The fetch in the prevailing wind directions, south-west and north-east, is about 200 m. The slope is almost flat with a 2 % downward gradient towards the south.

The soil, classified as a Calcic Kastanozem Skeletic (FAO 2008) and derived from calcium carbonate parent rock, boasts an alluvial origin and features a silt-clay texture.

2.2.5 Castellaro (IT-Cas)

The site is located in Lomellina, a typical rural area of the Po Valley in Northern Italy, in the municipality of Torre Beretti and Castellaro (Pavia) (45.07005° N; 8.71752° E, 88 m.a.s.l.). The field extends for 400 $\times$ 700 m and it has been cultivated with hybrid corn (*Zea mais* L., variety pioneer for silage production) in 2006 and with rice from 2007 to 2010. The rice cultivar was Balilla (japonica rice short grain), a typical management cycle starts with 25 cm deep ploughing in January, laser soil leveling in February, in March sugar beet industrial slops (2,250 kg ha^{-1}) and urea (75 kg ha^{-1}) are incorporated and compacted before sowing in April into water with 210 kg ha^{-1} of seed. Weed control in May, top dressing with Urea (90 kg ha^{-1}) in June. The average depth of standing water was maintained at about 5 cm in 2007 and 10 cm in 2008. The standing water level fluctuates during the flooded periods due to the continuous water flow through the rice field, a typical management practice for this crop. At the beginning of September the water is removed from the field and the harvest takes place in October. The final rice production was about 9 tree ha^{-1}. The soil originating from fluvial deposits is poorly drained and characterized by a ground water table very close to the surface. In most parts of the field it is a Calcic Gleysol (FAO 2008), with a loam to clay-loam texture, in other parts of the field we find Haplic and Mollic Gleysols. The A-horizon shows pH (H_2O) values ranging at 6.4–7.1, the organic carbon content is between 1.6 and 2.4 %. The gleyic color pattern of soil materials indicates dominant water saturation for most of the year in combination with poor drainage due to cultivation tillage and agricultural practices. During the observation period the average annual rainfall was 704 mm while the average annual temperature was 13.0 °C.

2.2.6 Castelvetrano (IT-Ctv)

The site is an olive orchard (cv. Nocellara del Belice) located in western Sicily at Castelvetrano (37.64416° N; 12.84638° E). The orchard has an extension of about 11 ha and has four different plots characterized by trees of different age (75 % 12–16 years old, 25 % 150 years old) but similar LAI (2.6–3.4) and trees height (3.4–3.7 m).

The soil texture is clay-loam with a volumetric water content at a field capacity of 36.1 and 17.4 % at −1.5 MPa soil water potential.

All of the plots were equipped with an irrigation system characterized by two micro sprinklers (8 liters $hour^{-1}$) per tree. During the vegetative season, the total volume of water varied between 1,500 and 2,000 m^3 ha depending on summer weather conditions. In accordance with the cultural practice adopted in this traditional olive-growing area, soil is clean cultivated during the vegetative season, whereas no weed control is undertaken during the rain season (autumn and winter).

The adopted tillage in the soil during the year allowed the growth of spontaneous annual herbaceous crops, characterized by the prevailing presence of grass (*Graminaceae* spp).

2.2.7 Castelporziano (IT-Cpz)

The experimental site was a 60 years-old coppice under conversion to high forest, inside the Presidential Reserve of Castelporziano (41.70525° N; 12.37611° E), 2.8 m.a.s.l., 25 km West of Rome. The natural Reserve covers about 4,800 ha, of which 85.6 % are forests. The climate is Mediterranean Type humid sub humid with an average yearly rainfall of 780 mm and an air temperature of 15.6 °C.

The forest structure is characterized by two layers: a dominant layer of *Quercus ilex* L. 12–15 m high and a shrub layer, 2–4 m high, with *Phillirea latifolia* L. *Pistacia lentiscus* L., *Erica arborea* L., *Cistus salvifolius* L.,*Cistus incanus* L., *Cytisus scoparius* L., and some vines (*Clematis flammula* L., *Hedera helix* L., *Rubia peregrina* L., *Smilax aspera* L.). Large trees over 100 years old, and up to 17 m high and 170 cm in diameter, represent only 1 % of the stand. LAI is in between 3.2 and 3.8. The herbaceous layers are represented by *Brachipodium sylvaticum* (Hudson) Beav., *Cyclamen repandum* Sibth et Sm., *Carex distachya* Desf., *Asperula laevigata* L., and *Alliaria petiolata* Bieb. The current *Q. ilex* forest represent a secondary succession after a fire destroyed the area in 1944. In 1985 the stand was converted from coppice into high forest.

The morphology of the reserve is mostly flat with altitudes ranging between 0 and 85 m.a.s.l. The soil is 40–60 cm deep with a pH 5.8 and the profile of the experiment is type A–C, with a fairly deep, well draining A horizon, rich in humus and calcium carbonate. The soil is sandy (75.2 % coarse sand, 11.2 % fine sand, 3.7 % coarse silt, 2.2 % fine silt, and 6.7 % clay). The soil organic matter is 3.4 % between 0 and 20 cm and 1.1 % between 20 and 40 cm. Organic carbon is 1.9 % in the organic horizon with a C/N ratio of 66 and is 0.65 % with a C/N ratio of 100 in other mineral Horizons.

The climate is characterized by a dry season extending for 4 months, from the beginning of May to the end of August. Water deficit during the dry seasons is 406 mm, while excess rainfall in other seasons is 410 mm. The scattered low summer rains are offset by high night humidity, which over a period of 10–15 days can contribute to 0.2–0.8 mm of precipitation.

2.2.8 Collelongo (IT-Col)

The experimental site is located near the village of Collelongo, Abruzzo region, in central Italy, close to the external belt of the Abruzzo, Lazio and Molise National Park. The Selva Piana forest stand (41.84936° N; 13.58814° E, 1,560 m.a.s.l.) belongs to a 3,000 ha forest community which is part of a wider forest area. The environmental and structural conditions of the stand are representative of central Apennines beech forests. The site was established in 1991 to study ecology and silviculture of Apenninian beech (*Fagus sylvatica* L.) forest. In 1993, the site was the first European forest to be instrumented to measure ecosystem level fluxes with Eddy Covariance (Valentini et al. 1996). In 1995–1996, the area became an ICP-Forests level II and an ICP-IM monitoring plot (http://icp-forests.net/). Since 2006, the site has become one of the main stations of the Long-Term Ecological Research site “Forests of the Apennine”. Over the years, the site was included in several EU and national research, monitoring and environmental projects. It is a pure beech forest with 830 trees ha^{-1}, with an average diameter of 22 cm and an average height of 21.5 m. The average tree age is approximately 120 years (2011). LAI ranges between 4.5 and 6.2 (1992–2011). Aboveground and belowground biomass is 171 Mg C ha^{-1}. The 40–100 cm deep soil is developed on calcareous bedrocks and is classified as humic alisol. Mineral soil carbon stock (0–80 cm) is 223 Mg C ha^{-1}, with 4.4 Mg C ha^{-1} stocked as litter (OL layer). The climate at the site is Mountain-Mediterranean with an average annual temperature of 6.9 °C and an average annual precipitation of 1,230 mm. Snow cover can last from mid December to mid March–early April.

2.2.9 La Mandria (IT-LMa)

The tower site (45.15258° N; 7.58259° E) is included into La Mandria regional park which has an extension of 3,000 ha and is located in the suburbs of Turin, near Venaria Reale, at about 350 m.a.s.l. The area is typical for peculiar geomorphic features originated by the presence of old terraces, residual of the ancient level of the main plain, which forms a hilly landscape with highlands and valleys, connected by steep slopes and characterized by regressive erosion of small rivers.

Vegetation is mainly constituted by the oak-hornbeam (*Quercus robur* L. and *Carpinus betulus* L.) of the high plains, which is considered an important forest-type to be studied and conserved. In the park there are also some plantations of red oaks (*Quercus rubra* L.), which cover around 20 % of the woods, some Robinia psuedoacacia formation on the slopes, some grasslands and poplar formations on the plains. Age of the oak-trees is about 80 years in average but an increase of trees mortality is registered along with wildlife high pressure which stops plant regeneration.

The soil on the top of the terraces is mainly 'paleosols', which according to the USDA classification are called Typic Fragiudalf, fine-silty, mixed, acid, mesic. One of the main features of the soil profile is connected to a temporary groundwater, placed at 60–70 cm below the ground level and a top-layer rich of silt, above a sub-layer called 'fragipan' which is very hard to root penetration. Therefore trees tend to stay with their root-apparatus within the upper part of the soil profile, often increasing the chances of tipping. Moreover, the greatest percentage of biological activity is detected in the topsoil, due to the difficulties of organic matter to deepen below the ground water zone.

Climate is classified as humid (second mesothermic with a low summer efficiency) with an annual rainfall of 1,030 mm and an annual temperature of 11.6 °C. The stand in the surrounding area of the flux tower has a measured biomass of 14 kg m^{-2}, a leaf litter of 1.4 kg m^{-2} and a tree density of 311 ha.

2.2.10 Lavarone (IT-Lav)

The Lavarone station is situated in the south east of the Trento province (45.95620° N; 11.28132° E; 1,300 m.a.s.l., average annual temperature 7.2 °C, total annual precipitation 1,150 mm) on a mountain plateau ranging from 1,000 to 2,000 m.a.s.l.

The whole study area covers nearly 80 km^2. This composite landscape is still widespread in the southern Alps, although it is undergoing a rapid transformation because of the combined effect of land use intensification in some more productive areas and the abandonment of more marginal ones. The landscape is characterized by the spatial succession of the three typical mountain management systems: grasslands (Arrenatheretum), pastures (Festucaetum) and mixed coniferous and broadleaf forests (Abietis-fagetum). The area analysed is characterized by an uneven-aged mixed forest dominated by *Abies alba* Mill. (70 %), *Fagus sylvatica* L. (15 %) and *Picea abies* (L.) H.Karst (15 %), with an average of 1,300 stems ha^{-1} (dbh > 7.5 cm) and a LAI of 9.6. The canopy has a dominant layer reaching 33–38 m and crown lower limits at about 12 m. In the understorey suppressed beeches form a discontinuous second layer from 0 to 4 m. The site is located on a gently rolling karst plateau and has a homogeneous vegetated fetch larger than 1 km in all directions, except for a 45° sector (300 m fetch in the SSW direction). Soils are generally well developed brown earth lying on a calcareous bedrock with the following characteristics in the 0–30 cm horizon: total SOC = 11.8 ± 1.8 kg C m^{-2}, total N = 0.61 ± 0.05 kg N m^{-2}, CN ratio = 15.9 ± 1.0.

2.2.11 Lecceto (IT-Lec)

Lecceto forest (43.30359° N; 11.26975° E) is a Holm oak coppice with a shift of 18–20 years located South-East of Siena, near the centre of the region of Tuscany, at about 300 m.a.s.l. It is the residual part of a broader forest named Selva del Lago for its proximity, during the Middle Ages, to a marshy area, subsequently drained in the XVIII century. That forest covered a surface of 10 km from north to south and 7 km from east to west. Because of the large use of wood during the Middle Ages, the Selva del Lago forest was subjected to specific rules of management and became a crucial crossroad between surrounding rural communities. In the XI century the most important Augustinian hermitage of Siena territory was founded in the middle of the forest. Since 1972 the hermitage was turned into a monastery of Augustinian cloistered nuns.

The forest covers an area of 900 ha and Holm oak represents 81 % of the total tree canopy; others species are represented by *Arbutus unedo* L., *Juniperus communis* L., *Quercus pubescens* L., *Phillyrea latifolia* L., *Fraxinus ornus* L.

The soil of the site is rocky and shallow, characterized by metamorphic schists and anagenite, and secondarily by cavernous limestone and carsism phenomena. The climate of Lecceto is a Mediterranean one, characterized by a strong seasonality with rainfall concentrated in autumn-winter, high inter-annual variability and frequent extreme events, such as summer drought and heatwaves. The average cumulated annual precipitation (1961–1990) at the nearest rain gauge is 780 mm, with a minimum in July (29 mm) and a maximum in November (108.7 mm); average annual temperature is 13.5 °C; July is the hottest month and January the coldest.

The stand in the surrounding area of the flux tower was coppiced about 15 years ago; its average stand height is around 9 m and the density of the stumps is about 1,200 ha^{-1}.

2.2.12 *Malga Arpaco (IT-Mal)*

Tesino Upland is located in the eastern part of Valsugana, in the southern part of external Dolomites; its Northern border is marked by the Lagorai mountains (2,700 m). The study site is located on a farm (Malga Arpaco) at 1,730 m.a.s.l. (46.11402° N; 11.70334° E), near Passo Brocon, it is a seasonality pasture (alpeggio) with pasture times from the beginnings of June to the end of August (around 100 days). The climate in the southern part of Dolomites is influenced by the Adriatic sea, while in the northen part it is more continental; the annual rainfall is about 1,100–1,550 mm concentrated in spring (May–June) and in autumn (October–November); the average temperature is lower than 0 °C from December to March.

The examined pasture is in the inferior alpine horizon. This vegetation layer is the upper limit of the microtherrmal Fagus forest and it may be constituted by Picea in the lower part and by Larix in the higher part. In this area, human activity changed the original vegetation to pastures and grasslands by cutting the forests, while in other sites the alpine pastures are typically above the upper tree vegetation line. The soil is an alfisol, sandy-loam and a Typic Hapludalfs (FAO 2008).

The Malga Arpaco pasture results to be sufficiently used, without bushes, pasturing paths, erosion; yet, the negative effects of understocking begin to appear, as evidenced by the existence of some limited overpastured areas surrounded by *Deschampsia* spp. and *Cirsium* spp.

From the floristic survey the grassland resulted to be composed predominantly by good fodder Poaceae (genera *Festuca*, *Poa*, *Phleum*), reduced by pasturing in summer; yet, the area shows, at least in some parts, the specific composition of a mid-damp and degraded grassland, with nitrophilous species, and the signs of a strong use, in transition from a grassland to an unrationally grazed pasture.

2.2.13 *Monte Bondone (IT-MBo)*

The study site was located at 1,550 m.a.s.l. on a mountain plateau in the Italian Alps (Viote del Monte Bondone, 46.01468° N; 11.04583° E). The average annual air temperature is 5.5 °C with monthly averages ranging from −2.7 °C in January to 14.4 °C in July. The average annual rainfall is 1,189 mm, with peaks in June (132 mm) and October (142 mm); snow cover occurs between November and April. The area is managed as an extensive meadow dominated by *Festuca rubra* L. (basal cover of 25 %), *Nardus stricta* L. (basal cover of 13 %) and *Trifolium* sp. (basal cover of 14.5 %), which represents a typical low productivity meadow of the alpine region. The maximum canopy height at the peak of the growing season (mid June to early July) can reach up to 30 cm. Meadows represent the main land use on this plateau (2 km^2) and are traditionally managed for hay production with low mineral fertilization and one cut per year in mid-July. Soil can be classified as a Typic Hapludalfs, lyme loamy, mixed, mesic with the following characteristics in the 0–30 cm horizon: total SOC = 9.4 ± 0.4 kg C m^{-2}, total N = 0.29 ± 0.02 kg N m^{-2}, soil bulk density = 0.79 ± 0.29 g cm^{-3}.

2.2.14 Negrisia (IT-Neg)

The site is located in North-Eastern Italy in a commercial vineyard located in Negrisia di Ponte di Piave (45.74756° N; 12.44673° E, 11 m.a.s.l.). The climate is mediterranean, with average annual precipitations of 833 mm and an average annual temperature of 13.1 °C. The station, set up in July 2005, is in the center of a one single 25 ha plot. The terrain is quite flat and both soil and canopy are very homogeneous and thus suitable for both micrometeorological and remote-sensing approaches. The vineyard is mainly composed of *Vitis vinifera* L. cv.'Carmenère N.' grafted on SO_4 rootstock. The vines were planted in 1992 in north–south oriented rows (500–600 m long) 2.50 m apart. Plant spacing on the rows is 1.30 m, resulting in a density of 3,076 vines per hectare. Vines were grown with a single trunk trained to spur-pruned cordon, at about 1.7 m aboveground, for vertical shaker harvesting. The floor is grass covered and a 1 m wide strip on the rows is chemically treated. The silty-clay loam soil (International textural classes), classified as Vertic Eutrudent (FAO 2008), is >1 m deep and highly homogeneous within the study site, rich in organic matter (about 18.2 g kg^{-1}) and with an average pH of 7.7. The vineyard receives standard management. Maximum canopy height was kept at 2.70 m, and no foliage could be found below the cordon. Maximum LAI, monitored both by direct and by indirect methods, ranges from 2.2 to 2.5.

2.2.15 Noè (IT-Noe)

The Noè site is located within a nature reserve called "Le Prigionette" (40.60613° N; 8.15146° E, 27 m.a.s.l.). The area is part of the Capo Caccia peninsula, in North-Western Sardinia, near the town of Alghero. The nature reserve, with a surface area of approximately 1,200 ha, is delimited by a cliff dropping to the sea on the north-western and western boundaries.

The climate is Mediterranean, semiarid with a warm summer, mild winter, and a prolonged water shortage from May through September. The averages for annual, minimum and maximum temperatures are respectively 15.9, 7.0, and 28.0 °C. The average annual thermal excursion is about 14 °C (10 °C in January and 24 °C in August). The average annual precipitation is 588 mm, mainly concentrated during the spring season. In the course of the autumn intense rainstorms often occur with high runoff and low water storage.

The main species are juniper (*Juniperus phoenicea* L.), lentisk (*Pistacia lentiscus* L.), tree phyllirea (*Phyllirea angustifolia* L.), and dwarf fan palm (*Chamaerops humilis* L.). These species form a sparsely vegetated shrub land, where juniper and lentisk, which respectively cover 53 and 22 % of the vegetated surface, are aggregated into variably-sized patches with bare ground in between. Phillyrea and palm can only be found as isolated elements inside the main patches. Other species typical of Mediterranean maquis are present on the experimental site: rosemary (*Rosmarinus officinalis* L.), *Genista corsica* (Loisel) DC., *Daphne gnidium* L., *Smilax aspera* L., *Euphorbia characias* L., *Helichrysum microphyllum* DC., *Asphodelus microcarpus* Salzm., and *Ferula communis* L. The vegetation is a secondary succession following a fire event occurred in 1963 and agricultural abandonment in 1970. Currently, this area has limited human activity. The average maquis height ranges between 0.93 and 1.43 m, and the ground cover varies between 42 and 91 %. Total LAI values range from 2.7 to 3.0. The soil is Lithic Xerorthent, 0.3–0.4 m deep, mainly composed of clay, and erosion is common.

2.2.16 Nonantola (IT-Non)

The Nonantola forest is a hardwood plantation located in a flat rural area of the Po valley in the Emilia Romagna region, close to Modena (44.69019° N; 11.09109° E 15 m.a.s.l.). The site was set up in 1992 with European fundings (Set-aside, Reg. CEE 797/85 and, being a re-forestation in a former rural area it is a proper representation of a Kyoto forest—Sect. 3.3 of the Kyoto Protocol). The forest is still growing and has not yet reached the final stage, its dominant species consisting of: Oak (*Quercus robur* L.) 35 %, Ash (*Fraxinus* spp.) 25 %, Maple (*Acer campestre* L.) 12 %, Willow (*Salix alba* L.) 8 %, Poplar (*Populus alba* L.) 6 %, Cerry (*Prunus mahaleb* L.), other 8 %. The total area is 38.6 ha with a tree density of 1,100 trees ha^{-1} at the plantation stage, and reduced to 826 trees ha^{-1} 10 years after. The soil, by Soil USDA Taxonomy is Fine, Mixed, Mesic Entic Chromusterts with a 60 % of clay, and a soil organic matter (SOM) of 49.8 Mg C ha^{-1}estimated in 1992 and 53.7 Mg C ha^{-1} measured in 2000. The average annual temperature of 13.1 °C and the average annual rainfall of 719 mm. characterize the site as located in Mediterranean climate, with rainfall concentrated in autumn-winter, summer drought and some heat waves events. Fluxes have been measured since 2001 by means of the Eddy Covariance system placed at 13 m above the forest floor and steadily to rise during tree increase.

2.2.17 Parco Ticino (IT-PT1)

The study site is located at 60 m.a.s.l., about 10 km North-West of the city of Pavia (45.20087° N; 9.06104° E) in the municipal area of Zerbolò, within the Ticino Natural Park in Northern Italy. Poplar land-use was introduced in the area after 1973 when a fire destroyed the foregoing natural land cover, a floodplain forest with dominant oak

(*Quercus robur* L.), ash (*Fraxinus excelsior* L.), white poplar (*Populus alba* L.) and black alder (*Alnus glutinosa* (L.) Gaertn). A relict of the natural forest is represented by the 200 year-old Bosco Siro Negri located within 1 km. The poplar high stand of 46 ha (Populus × canadensis Moench, Clone I-214) was 12 years old in 2002, the spacing was 6 × 6 m and the tree density was 278 tree ha^{-1}. The average tree height was 26.6 m, while the average diameter at breast height (dbh) was 0.33 m and the stem basal area was 20.5 m^2 ha^{-1} in 2005, when the site was logged. In order to carry out replanting after logging, residues were removed, stumps were drilled to allow a 60 cm deep ploughing, and 4 m long shoots were inserted for about 150 cm into the ground. In general, management intensity was low with removal of ground vegetation (mainly Artemisia and Poa) by harrowing during early summer and occasional irrigation on demand. During 2002–2005 one fertilization event of 300 kg ha^{-1} of urea occurred in May 2002, and two natural flooding events in November 2002 and October 2004. Soils are Arenosols and Regosols not well developed on recent alluvial sand and gravel deposits of the Ticino river. Topsoil shows a bulk density of 0.9–1.4 g cm^{-3}, the nitrogen and carbon content are 5.23 kg C m^{-2} and 1.36 kg N m^{-2}, pH (H_2O) ranges at 5.5–7.0. The soil texture is sandy-loam (60.4 % sand, 30 % silt and 9.6 % clay). The climate of the site is classified temperate continental, with an average annual rainfall of 912 mm and an average temperature of 12.5 °C. Budburst usually occurs at the beginning of April and senescence in October, with the LAI peaking at around 2.0 in July. Winds show a prevailing direction from the north-east and south-west. Atmospheric nitrogen deposition is about 20 kg N ha^{-1} per year.

2.2.18 Pianosa (IT-Pia)

The Island of Pianosa (42.58437° N; 10.07804° E) is one of the seven islands of the Tuscan Archipelago with a total area of 10.2 km^2 and a coastal perimeter of approximately 20 km. The island is almost completely flat, with some small undulations, its highest elevation is 29 m.a.s.l., while the average is about 18 m.a.s.l. On the basis of a historical meteorological dataset (1951–2009), the average air temperature is 15.8 °C and the average annual rainfall is 497 mm, ranging between a minimum of 176 mm (1999) and a maximum of 716.2 mm (1984). A clear seasonal precipitation pattern shows a maximum from October to December followed by a decrease with a minimum value in July. The soils are sandy-loam or sand, alkaline, rich in carbonates, with a moderate content of rock fragments, mainly classifiable as Leptosols. Three main ecosystems have been mapped: abandoned crops and pastures, Mediterranean macchia and woodlands. Abandoned croplands and pastures are the main ecosystems in terms of extent (52 %) but are less representative in terms of total biomass (26 %). The land previously used for agriculture is now covered by a species association typical of degraded Mediterranean agricultural soils. The Mediterranean macchia vegetation exists at different evolutionary stages as a consequence of the progressive re-naturalization process which the island is currently undergoing. The Island of Pianosa represents an interesting case study to assess the evolution related to land use change in a Mediterranean environment, since the island was intensively used for agricultural production presumably since the Romans. In particular, intensive soil cultivation was carried out by the prisoners of the local Penal Colony in the last century and subsequently abandoned at the end of the 1990s.

2.2.19 Roccarespampani Forest (IT-Ro1 and IT-Ro2)

The two sites, along with the IT-Ro3 and IT-Ro4 described below, are part of the Roccarespampani cluster within a publicly owned farm in the Province of Viterbo. The climate of the sites is typical Mediterranean, with dry and hot summers and relatively mild and wet winters. The average annual temperature is of 15 °C with a total rainfall of 800 mm. The precipitation distribution pattern typically features a drought period in in the months of July and August. The bedrock is of volcanic origin and the soil is a 90 cm deep Luvisol (Tedeschi et al. 2006).

The Roccarespampani forest stretches for about 1,300 ha in a fairly flat area varying from 120 to 160 m.a.s.l. and it is mostly composed of Turkey oak (*Quercus cerris* L.) stands, historically managed under the silvicultural system of coppice-with-standards in the last 200 years with a rotation cycle ranging 15–20 years and it has been arranged in 15 sequentially aged compartments with an average area of 85 ha.

Eddy Covariance fluxes were monitored over two sites established in distinct compartments. The IT-Ro1 site (42.40812° N; 11.93001° E) operated in the period 2000–2009 and was representative for the initial developmental stages (0–9 years) of the forest after the forest cut (coppicing), which occurred in December 1999. The IT-Ro2 site (42.39026° N, 11.92093° E) is instead representative of older stages (11 years) and it was set up in 2002. In terms of vegetation composition and structure, the earlier stages are characterized by a dense regrowth of shoots (~6,000 trees ha^{-1}) on coppiced stumps accompanied by shrubs such as *Cytisus scoparius* (L.) Link, *Spartium junceum* L., *Crataegus* spp., *Paliurus spina-christi* Mill., *Rosa canina* L., while oak standard trees in the range of 20–60 years (~100 trees ha^{-1}) tower over the low canopy. Later stages exhibit a less dense structure (~3,000 trees ha^{-1}) with a 16–20 m high main tree canopy, largely made up of *Q. cerris*, which includes also *Quercus pubescens* Willd., *Fraxinus ornus* L., *Ulmus minor* Mill., *Ostrya carpinifolia* Scop., *Acer monspessulanum* L. and *A. campestre* L. An understorey layer typically composed of *Phillyrea latifolia* L., *Crataegus* spp., *Ruscus aculeatus* L., *Cornus* spp., *Prunus spinosa* L., *Sorbus* spp. completes the forest vegetation (PGAF 2009).

2.2.20 Roccarespampani Crops (IT-Ro3 and IT-Ro4)

The sites are part of the Roccarespampani cluster (see Sect. 2.2.19) and were established in the late summer of 2007 in a flat, non-irrigated cropland. The two Eddy Covariance systems were installed in two adjacent arable fields (42.37539° N;

11.91542° E for IT-Ro3 and 42.37333° N; 11.91922° E for IT-Ro4). The two cropland fields are of about 15 ha each allowing a minimum distance of 110 m from the fields' edges, with surrounding fields being similarly covered by non-irrigated cropland. The soil is an Eutric Cambisol with a clay loam texture (40 % clay, 30 % silt, 30 % sand) and, on average, 10.8 g kg^{-1} of organic carbon and 1.3 g kg^{-1} of total nitrogen in the cultivated layer (0–30 cm).

The fields have been under dry arable cultivation for the last two decades, under a rotation including primarily durum wheat (*Triticum durum* Desf.) and annual forage crops, mainly crimson clover (*Trifolium incarnatum*, L.) as a pure stand or as a mixture with oats (*Avena sativa* L.) and common vetch (*Vicia sativa* L.). In the 2007–2008 and 2008–2009 growth seasons, the two sites were employed for a paired comparison of different agronomic management for biofuel crop productions, namely a winter crop, i.e. rapeseed (*Brassica napa* L.), in 2008 and a spring crop, i.e. sunflower (*Heliantus annuus*, L.), in 2009. One of the fields (IT-Ro3) was managed conventionally using mouldboard ploughing and disk harrowing, while the other (IT-Ro4) was managed as a no-tillage system. After 2009, the farm was converted into an organic agriculture regime and the no-tillage field, typically requiring chemical weeding, was then reverted to conventional tillage. Both fields were occupied by organically grown crimson clover (*T. incarnatum*) for seed production in the 2009–2010 growth season. In 2010–2011, the IT-Ro3 field was employed for (organic) durum wheat production, whereas the IT-Ro4 field was occupied by crimson clover employed as an annual forage crop, i.e. occasionally grazed by cattle. In the 2011–2012 growth season one of the fields (IT-Ro4) was managed as crimson clover for seed and straw production while the other (IT-Ro3) was occupied by a mixture of vetch and oats employed as a forage crop with occasional cattle grazing and final hay harvest in June.

2.2.21 Renon (IT-Ren)

The Renon-Selva Verde site (46.58686° N; 11.43369° E, elevation about 1,735 m.a.s.l.) is located in the municipality of Renon, at a distance of 12.2 km North-Northeast from the town of Bolzano. Eddy Covariance measurements started in the year 1997. The site is placed on a porphyric plateau; the soil is classified as Haplic Podsol following FAO (2008).

The site vegetation, a subalpine coniferous forest, is of natural origin and is used for wood production. As a result of the traditional harvesting method, which consists of irregular cuttings of 50–80 cubic meters, overall the forest is unevenly aged, but with homogenous groups. The largest group present in the area has been approximately growing since the year 1820, after the Napoleonic wars. The main forest species is spruce (*Picea abies* (L.) Karst., 85 % in number) followed by cembran pine (*Pinus cembra* L., 12 %) and larch (*Larix decidua* Mill., 3 %). In the clearings, covering approximately 15 % of the area, the dominant grass species is *Deschampsia flexuosa* L. The canopy is irregular, with a maximal height of 29 m. The average LAI, measured by hemispherical photographs, is 5.1 (Montagnani et al. 2009).

The climate is strongly influenced by elevation, with cool summers and moderately cold winters. During the last decade, the average temperature was 6.21 °C and the annual precipitation was 964.21 mm (Niu et al. 2011).

2.2.22 San Rossore (IT-SRo)

The study site is an even aged Pinus pinaster forest located 9 km west of Pisa (central Italy) and 800 m east of seashore inside the regional park of San Rossore-Migliarino-Massaciuccoli, (43.72786° N; 10.28444° E, elevation 6 m.a.s.l.) The park extends for 24,000 ha about 30 km between Livorno and Viareggio along the Tyrrhenian coast with an estuarial morphology characterized by the presence of alluvial deposits and sandy dunes between the rivers Serchio and Arno. *Pinus pinaster* Aiton was introduced in 1771 in a strip of ca 10 × 1 km behind the dunes in order to protect the economically important *P. pinea* L. plantations more inland. Before human intervention, the *P. pinaster* area was covered by Mediterranean sclerophyllous forests and scrubs, a formation of Ligurian-Thyrrenian meso-Mediterranean holm oak forests (*Quercus ilex* L.) with *Fraxinus ornus* L., *Ostrya carpinifolia* Scop., *Viburnum tinus* L. The current *P. pinaster* forest established from seed after a big fire in 1944 and was never thinned systematically. A survey of the experimental plot in 2000 showed a tree density of 587 trees per ha (84 % *P. pinaster*, 12 % *P. pinea*, 4 % *Q. ilex* in the understorey), a basal area of 40 m^2 ha^{-1}, an average canopy height of 18.7 m, a LAI of 2.8 and a total aboveground and root biomass of 9.7 and 1.9 kg C m^{-2}. Due to a very high ungulate density, there is no ground vegetation and only few shrubs like *Phyllirea angustifolia* L., *Myrtus communis* L., *Erica arborea* L. The soil in the pinewood is a regosol with a 93 % sand texture extending down to 4 m depth. The top 10 cm of soil is characterized by pH (H_2O) of 6.5 and C/N ratio of 25.2; rooting depth is limited by the water table varying between 100 cm in winter and 200 cm in summer. The climate of San Rossore is Mediterranean sub-humid, with an average annual rainfall of 876 mm and an annual temperature of 15.6 °C. The wind regime is characterized by a land–sea breeze circulation with air flows from the west during the day and from the east during the night. A large scale dye back of *P. pinaster* took place in the 70s due to the combined impact of marine aerosol and surfactants. Currently, the *P. pinaster* area is under a large scale attack of the maritime pine bast scale (*Matsucoccus feytaudi*) and is going to be transformed into a more natural land cover with dominant *Q.ilex*.

2.2.23 Tolfa (IT-Tol)

The Tolfa-Allumiere site (42.18967° N; 11.92155° E) is an evergreen broadleaf managed forest extending for about 6 ha, primarily characterized as Mediterranean Macchia and located in a small and isolated mountainous chain in Central Italy, close to the Tyrrhenian Sea. The elevation is 422 m.a.s.l. and the climate is essentially Mediterranean with a small increase in precipitation during autumn and spring, essentially driven by local meteorological circulation and updraft. The average annual precipitation is 650 mm while the average annual temperature is 15 °C. The site has been set up as part of a large scale manipulation experiment called LASMEX in the framework of EU-funded MIND project which aimed at studying the effect of increased or decreased precipitation intensities on Mediterranean ecosystems through rain exclusion or irrigation. This specific site is the reference site with unmodified precipitation regime. The vegetation is a coppice rotation primarily composed by *Arbutus unedo* L. (65 %) and *Erica arborea* L. (13 %) and secondarily by *Fraxinus ornus* L. and *Quercus ilex* L. The entire canopy grew up after fire events occurred on site in 1984. The soil is of volcanic origin and classified as Andosols, with an average depth of around 30 cm and acid (pH = 4).

2.2.24 Torgnon (IT-Tor)

The site, active since June 2008, is an old grazed pasture abandoned in late 1990s in the northwestern Italian Alps. It is located a few kilometers away from the village of Torgnon in the Aosta Valley region at an elevation of 2,160 m.a.s.l. (45.84444° N; 7.57806° E).

The dominant vegetation is composed by *Nardus stricta* L., 35 %, *Festuca nigrescens* All. 11 %, *Arnica montana* L. 8 %, *Carex sempervirens* Vill. 5 %, *Geum montanum* L. 5 %, *Anthoxanthum alpinum* L.L. 4 %, *Potentilla aurea* L. 4 %, *Trifolium alpinum* L. 4 %. Considering canopy properties, maximum seasonal LAI is about 2.9 and average canopy height is 18 cm. The terrain slopes gently (4 °) and the soil is classified as Cambisol (WRB classification), with the following characteristics in the 0–20 cm horizon: organic C = 2.8 kg C m^{-2} and total N = 0.22 kg N m^{-2}.

The site is characterized by an intra-alpine semi-continental climate, with strong seasonality, the average annual temperature is of 3.1 °C and the average annual precipitation is of about 880 mm. On average, from the end of October to late May, the site is covered by a thick snow mantle (90–120 cm) which limits the growing season length to four-five months. Growing season cumulative precipitation can show huge variations (from 160 to 630 mm) and the average daily air temperature varies between 4 and 13 °C.

2.2.25 Valle dell'Adige (IT-VdA)

The Valle dell'Adige site is located in a vineyard in the Valle dell'Adige (46.19678° N; 11.11354° E; Mezzolombardo) at an altitude of 206 m.a.s.l. The average annual air temperature is 11.8 °C while the average annual precipitation is 822 mm. The soil was classified as a Gleyic/Haplic Fluvisol (FAO 2008) with the following characteristics in the 0–30 cm horizon: sand 42 %, silt 47 %, clay

11 %, total soil organic content (SOC) = 5.85 kg C m^{-2}, total N = 0.66 kg N m^{-2}, bulk density 0.96 g soil cm^{-3}. The vineyard covers an area of approximately 2.5 × 1 km cultivated with Vitis vinifera L. cultivar "Teroldego rotaliano", and includes small patches of apple trees. The vegetation is about 2.0 m tall with more than 50 % ground shading. Mass and energy fluxes have been continuously measured since 2008 at the site with an Eddy Covariance system mounted at 7 m height.

The growing season starts in the middle of March and lasts until the end of September with maximum carbon dioxide uptake at the end of April (for high radiation and water availability) and in July. During the winter season (from October to March) the vineyard acted as a carbon source.

2.2.26 Vigevano (IT-Vig)

The experimental site is located near the town of Vigevano (45.29591° N; 8.87554° E, 117 m.a.s.l.) in the Lombardy Region, in the Province of Pavia and it consists of a poplar short rotation coppice (SRC) (*P. generosa* × *P. nigra* clone Pegaso) of about 80 ha. The traditional agricultural land use of the area is a corn-rice rotation transformed into SRC in March 2004 by planting one year old seedlings in a double row design with a spacing of 2.8 × 0.75 × 0.45 m corresponding to a density of 12.500 plants ha^{-1}. Before planting, soil was ploughed and disk-harrowed. After planting, a pre-emergent herbicide was applied at 2 kg ha^{-1} with a pull type sprayer. In the following years weed control was done mechanically four times in spring and summer with a rototiller. From May to September the experimental field was irrigated with a volume of water of 1,500 m^{-3} ha^{-1}. Poplar shoots were harvested and chipped annually during winter with a special harvester and used in a nearby incinerator for energy production. No fertilizer was applied during the first three years of SRC cultivation. The soil is from fluvial sediments and has a sandy loam texture (53 % sand, 30 % silt and 17 % clay), bulk density is 1.21 g cm^{-3}, pH (H_2O) is 6.5 in the uppermost 32 cm. The net primary production (NPP) as measured as sum of stem+branch, coarse roots, fine roots, stump, leaf litter and fine root litter increased in 2004–2006 from 0.69 to 1.05 to 1.3 kg C m^{-2} $year^{-1}$. The height of the canopy reached 4.5 m in one year, the LAI peaked in July-August at 4.2. The average specific leaf area (SLA) of poplar leaves was 296.56 (±60.09) cm^2 g DM^{-1}. The climate of the site is classified as Humid Subtropical-Mid Mild Latitude (Cfa)—Koppen Climate Classification, with an average annual rainfall of around 1,000 mm and an average temperature of about 12.5 °C.

2.3 Example of Measurements

The NEE measurements using the Eddy Covariance technique are typically half hourly average values expressed in μmol CO_2 m^{-2} s^{-1}. The data can be clearly used to calculate daily to annual sums and budgets which are then expressed in

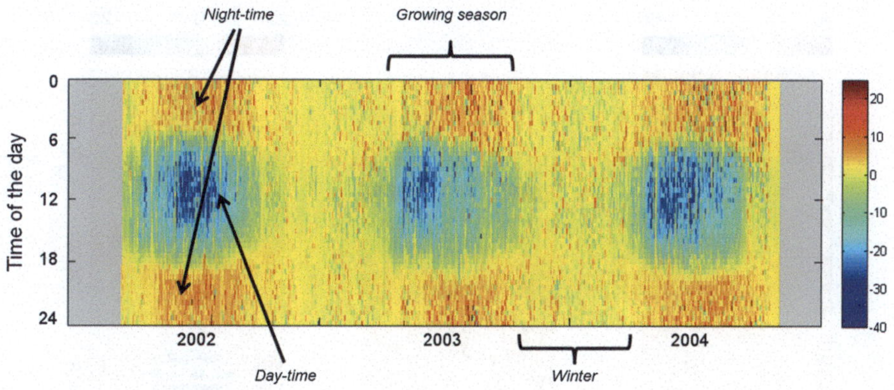

Fig. 2.2 Example of fingerprint plot to visualize Eddy Covariance time series. The *colors* are instantaneous (half hour) CO_2 fluxes in μmol CO_2 m^{-2} s^{-1} (site IT-PT1). It is possible to see clearly the seasonal dynamic with emissions of CO_2 during winter when only respiration is present and the diurnal cycle with CO_2 assimilation in the daytime. The minor CO_2 uptake in the summer of 2003 is to be noted, as a consequence of the European heat wave (Ciais et al. 2005)

general as g C m^{-2} day^{-1}. By convention, negative values are fluxes going from the atmosphere to the ecosystem (carbon sink).

One of the best ways in order to have a graphical overview of the measurements is the "fingerprint" plot (Fig. 2.2): it is an image where on the X axis there are the days, on the Y axis the hours and the color indicate the flux magnitude and sign. Using this type of plot it is possible to easily follow and analyze the diurnal, seasonal and inter annual patterns and identify anomalies.

The Fig. 2.3 shows the comparison of two cropland sites (IT-BCi cultivated with maize all the years and IT-Be2 cultivated with alfalfa in 2007–2009 and maize in 2010) and it is possible to see the effect of the management options: tillage before the carbon uptake period (CUP) in IT-BCi, the four harvest in 2007–2009 in IT-Be2 and the difference in the magnitude of the uptake, with the maize reaching higher values (C4 photosynthesis) in IT-BCi and also in IT-Be2 in 2010 (see Sect. 2.2.2). In the same figure a grassland site with a long time series of data is also reported (IT-MBo) and it is possible to identify the differences in the seasonal cycle with a winter basically without activities because of the snow and cold temperature.

Another interesting example is given by the comparison of the two oak coppice forests in central Italy (Fig. 2.4). The two sites are in two compartments with different ages with IT-Ro1 monitoring a forest just after the harvesting occurred in 2000 and IT-Ro2 in an older forest now ready to be harvested. It is interesting to note that the fluxes are higher in the mature forest but also the young stand recovered quite rapidly a good photosynthesis capacity. In addition in both the stands it is possible to see the effect of the heat-wave in 2003 (Ciais et al. 2005) with a strong reduction of the carbon assimilation in summer and a similar anomaly also in 2008.

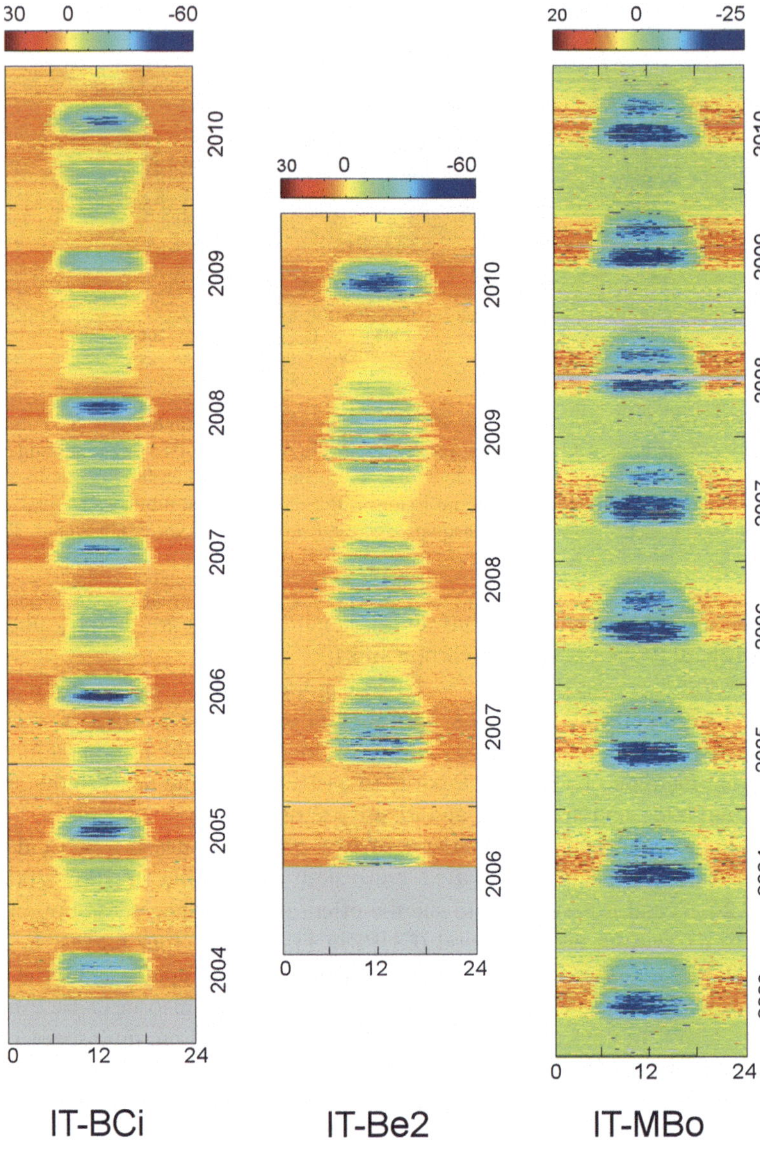

Fig. 2.3 Comparison of two cropland sites: IT-BCi cultivated with maize all the years and IT-Be2 cultivated with alfalfa in 2007–2009 and maize in 2010. The *colors* are instantaneous (half hour) CO_2 fluxes in umol CO_2 m^{-2} s^{-1}. It is possible to note the different magnitude of the uptake between the two species and the effect of the management practices. The third site (IT-MBo) is a grassland site and it is interesting to note the low fluxes during winter and the growing season characterized by two assimilation picks

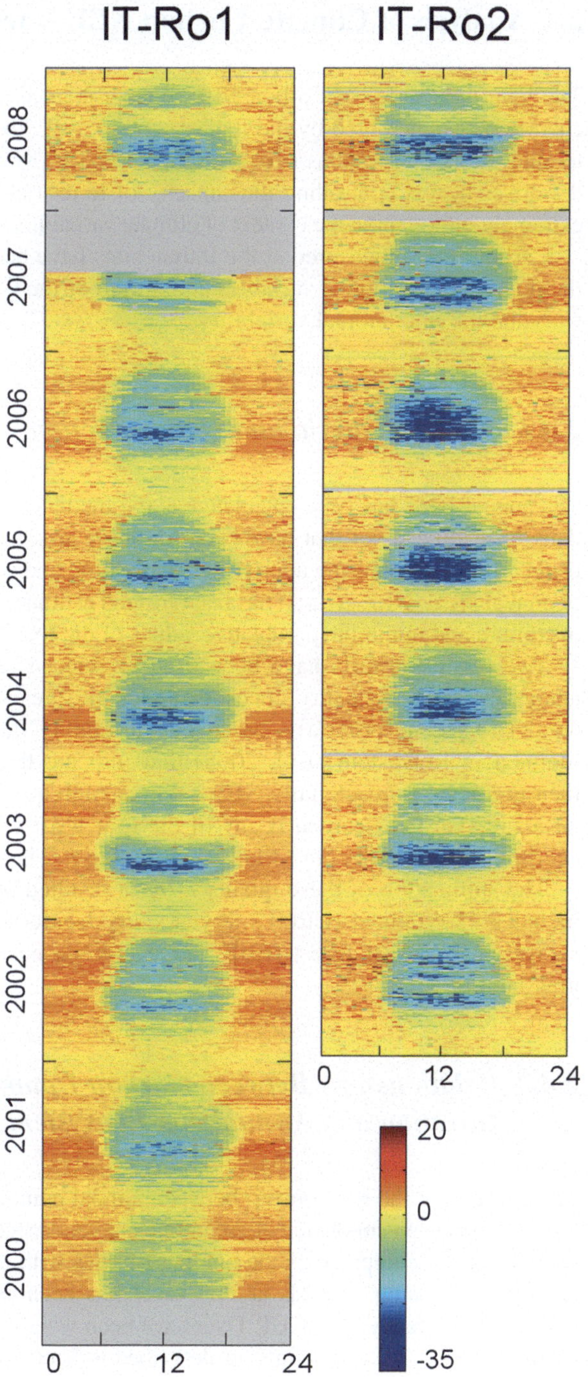

Fig. 2.4 Comparison of the two forest sites in Roccarespampani: IT-Ro1 and IT-Ro2 are in two different compartments with different stand ages but it is possible to identify common patterns (2003 and 2008) and seasonal dynamics. In addition to this, it is possible to see in IT-Ro1 the recovery of the forest after the harvesting in 2000. The *colors* are instantaneous (half hour) CO_2 fluxes in μmol CO_2 m^{-2} s^{-1}

2.4 Analysis of Climate-Carbon NEE Interactions

The long time series of Carbon NEE, meteorological variables and ecosystem status characteristics measured by the Italian network of Eddy Covariance sites can be used to analyze the interactions between climate and ecosystem processes. Understanding the driver factors controlling the capacity of terrestrial ecosystems to store carbon is extremely important in the context of climate variability and climate changes processes.

Carbon fluxes measured at the Italian sites have been used to analyze the role of the main meteorological variables and length of the CUP in the annual NEE across sites and the inter annual variability.

2.4.1 Quality Criteria and Data Selection

The annual NEE has been calculated as the median of the four cumulative values of daily NEE time series that have been routinely produced in the Carbo Italy database using different processing approaches (gap filling and storage correction). Uncertainty of annual NEE ($NEE_{max-min}$) has been instead estimated as the difference between the maximum and minimum cumulative values of the four daily NEE time series.

Two quality criteria have been used to select site years to be included in the analysis. The first quality check excludes site years with less than 330 days of data. The second criteria is based on the quality flags of daily NEE time series (defined in Reichstein et al. 2005) that indicate the percentage of high original measurements for each daily NEE value or high quality gap filled data: in particular a site year is discarded if any of the four NEE versions has a within year percentage of low quality days (daily quality flag < 0.85) higher than 25 %.

The application of these quality criteria resulted in a dataset for analysis which includes 22 cropland (CRO) site years, 26 deciduous broadleaf forest (DBF), 18 evergreen broadleaf forest (EBF), 27 evergreen needle leaf forest (ENF) and 14 grassland (GRA).

2.4.2 Relationships Between Carbon Uptake Period Duration and Annual Net Ecosystem Exchange

The analysis of the relationship between annual cumulated NEE and the duration of the CUP has been conducted aggregating site years according to PFT. CUP for all site years has been computed as the number of days with daily NEE < -1 g C m^{-2} d^{-1}. A linear regression analysis at PFT level has been conducted in order to evaluate annual NEE sensitivity to CUP. Data have been weighted according to the uncertainty estimates ($NEE_{max-min}$) defined as described in Sect. 2.4.1.

Results of the linear regression analysis between annual NEE and CUP are reported in Fig. 2.5. Higher R^2 values are obtained for EBF, ENF and GRA as presented in Fig. 2.6. DBF is the PFT with the lowest R^2 value probably because

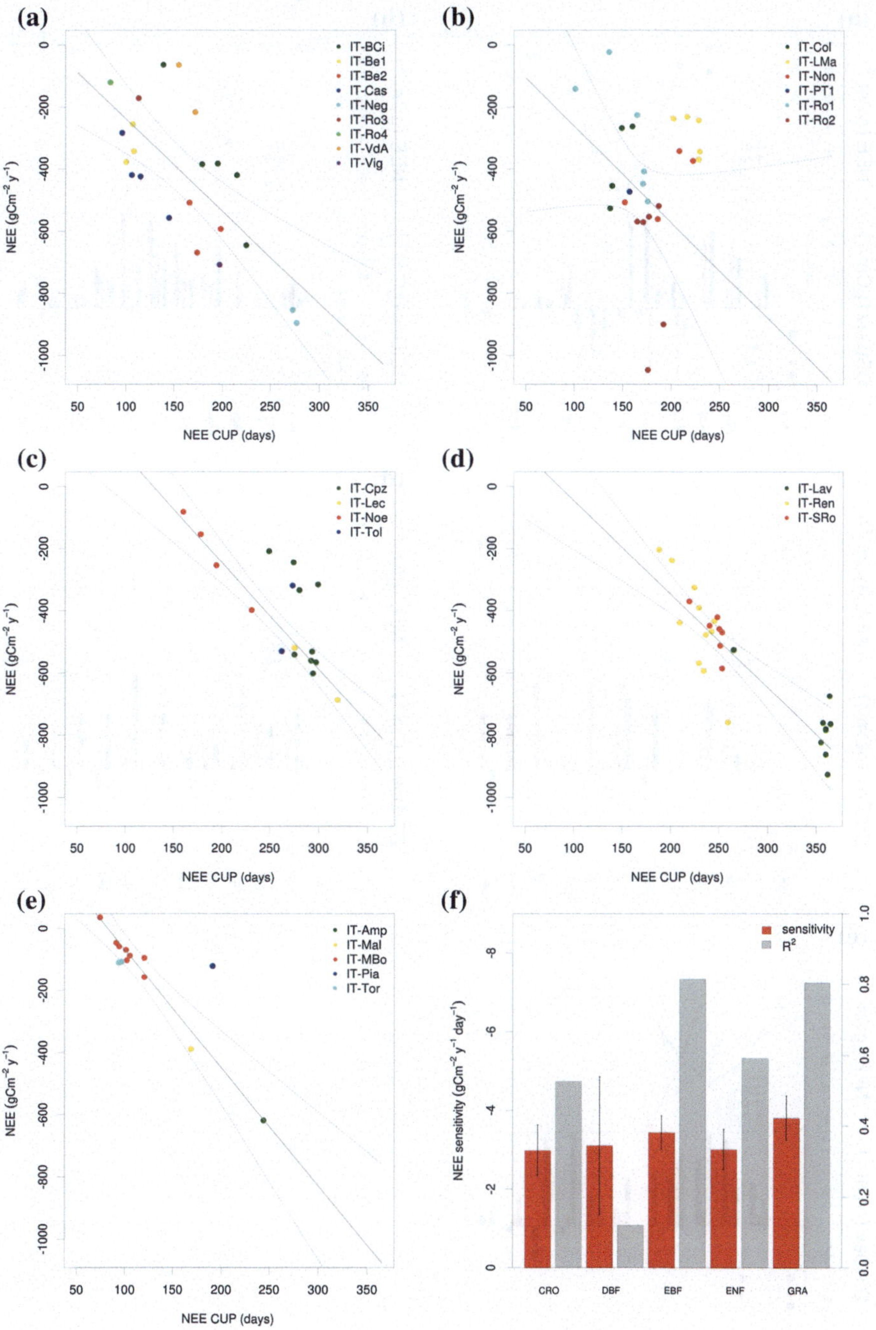

Fig. 2.5 Linear regression analysis between NEE annual values and CUP for each PFT. Panel **f** presents the sensitivity (*red bars* with s.e.) and R^2 (*grey bars*) of the relation between NEE and CUP for each PFT

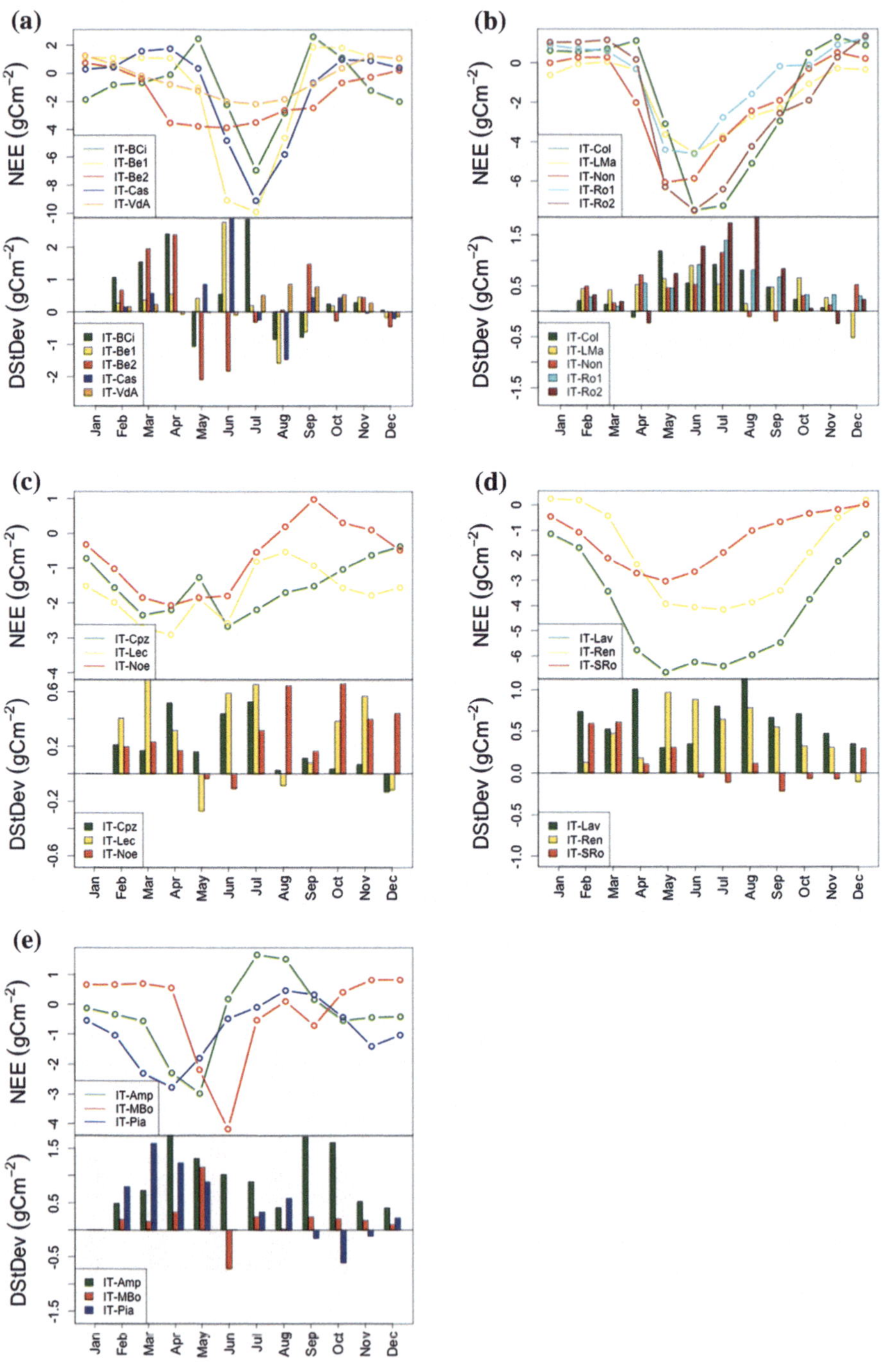

Fig. 2.6 Time series of average monthly NEE for the period reported in the database for each site (*upper panel*); barplot of DStDev over the observation period (*lower panel*). Different bars represent different sites. Different plots represents different PFTs

of the large variability of ecosystems traits of sites that are included in this PFT; in fact the DBF sites in the networks are highly heterogeneous, with mature forests (IT-Col), coppices with different ages (IT-Ro1 and IT-Ro2) and a poplar plantation (IT-PT1). PFT annual NEE sensitivity to CUP variations range between -2.97 ± 0.64 for CRO and -3.80 ± 0.56 C m^{-2} yr^{-1} day^{-1} for GRA.

Within some PFTs, sites with particular behaviors emerged (e.g. IT-Cpz for EBF and IT-Pia, included in this analysis in the grassland PFT) probably because of the effect of some site properties (in IT-Cpz for example there is the effect of the water table depth) or ecological parameters (e.g. LAI, fertility, …) on the relation between annual NEE and CUP. These effects generally resulted in a similar NEE sensitivity to CUP variations but with different absolute values.

Moreover CRO and GRA, PFTs where management can play an important role, showed quite good relationship between NEE and CUP and sensitivity values similar to those obtained in the other PFTs (Fig. 2.5a, e).

2.4.3 Analysis of the Relationship Between Annual NEE and Climate Variables

The analysis of the relationship between annual cumulated NEE and climate variables has been conducted by aggregating site years according to the PTF. The analysis has been conducted in three different steps:

1. As a first attempt, a correlation analysis between annual NEE and average annual climate variables has been conducted. As climate variables temperature, precipitation, radiation and vapor pressure deficit (VPD) have been used.
2. In a second step, the stepwise regression has been used for the selection of the best predictive variables of annual NEE. The predictive variables tested were meteorological variables (T, Prec, VPD, Rg), LAI- and phenological indicators (i.e. CUP). A stepwise approach based on the Akaike's Information Criterion (AIC) for model selection has been implemented. The AIC is a measure of the trade-off between the goodness-of-fit (model explanatory power) and model complexity (number of parameters). Therefore, the stepwise based on AIC is a multiple regression method for variable selection which accounts on one hand for the model explanatory power, and on the other for the increasing complexity of the model when additional variables are tested in the multiple regression model (Venables and Ripley 2002; Yamashita et al. 2007). The stepwise AIC was preferred to other stepwise methods for variable selection since it can be applied to non-normally distributed data (Yamashita et al. 2007). The data have been weighted according to the uncertainty estimates ($NEE_{max\text{-}min}$), defined as described in Sect. 2.4.1 The outcome of the stepwise AIC is a multiple linear model including the best set of predictive variables explaining the variability of annual NEE. Therefore, by applying stepwise AIC as explained above, the drivers explaining the variability of NEE for each PFT have been identified while the coefficients estimated for each selected variable represent the sensitivity of NEE to variations of each predictor.

3. In the third analysis, the inter-annual variability has been analyzed defining for each site with more than 3 years of data, the phenological period that mostly contribute to annual NEE variability. Following Marcolla et al. (2011), for each site-year the time series of cumulative monthly NEE and its standard deviation have been computed. According to Marcolla et al. (2011), variations of standard deviation from one month to the preceeding one (DStDev) quantifies the contribution of each month to the observed interannual variability (IAV). The analysis has been conducted also at weekly time scale with comparable results (data not shown). Resulting DStDev positive values means that the specific period contributes to the increase of the between year flux variability. Positive peaks of DStDev mean that climatic conditions in the period in which the peak occurs largely influence the inter annual variability. Periods with negative values of DStDev tend instead to mitigate the inter annual variability of NEE. Once the critical period for each PFT has been identified, the stepwise AIC regression between annual NEE and different predictors has been re-computed. Not only have the average annual values of climate predictors been tested, but also sub-annual variable aggregation (i.e. average temperature in spring and summer; cumulative precipitation in summer etc.). For each PFT the main driver controlling spatial and year-to-year variation in NEE has then been identified.

The results of the pair wise correlation analysis between annual NEE and climate variables conducted for each PFT are reported in Table 2.1. The results show that average annual climatic variables cannot explain the variability of NEE except for the EBF, for which a statistically significant correlation ($p < 0.05$) with Ta, Ts and Rg have been found and for ENF, for which a correlation between NEE and Soil Water Content (SWC) is observed.

In Table 2.2 the results of the stepwise AIC regression between cumulative NEE and annual climatic predictors are reported. Even including multiple climate predictors, a statistical significant correlation with NEE is observed only for the evergreen PFTs (EBF and ENF).

By including as predictors the CUP and LAI, an important improvement of the R^2adj has been observed, highlighting the importance of structural site

Table 2.1 Pairwise correlation (r) between annual NEE cumulated and annual average of meteorological variables

	Rg	Ta	Ts	VPD	Precip	SWC
ENF	0.01	0.00	0.00	0.00	0.00	0.44
DBF	0.02	0.00	0.01	0.01	0.03	0.02
EBF	0.49	0.70	0.58	0.01	0.08	0.04
CRO	0.03	0.00	0.03	0.09	0.04	0.02
GRA	0.26	0.00	0.02	0.01	0.00	0.21

Bold numbers represent statistical significant correlations ($p < 0.01$). *Rg* Shortwave Incoming Radiation, *Ta* Air Temperature, *Ts* Soil Temperature, *VPD* Vapur Pressure Deficit, *Precip* Precipitation, *SWC* Soil Water Content

Table 2.2 Results of model selection conducted with the stepwise AIC method using as predictor annual averages of meteorological variables

PFT	Ta	Precip	Intercept	R^2adj	*P*	AIC	N
ENF	13.08 (8.52)	−0.33 (0.15)	−456.3	0.36	<0.001	108.6	12
DBF					NS	NS	14
EBF	95.04 (25.02)	−0.16 (0.10)	−1,783.59	0.66	<0.01	88.99	12
CRO					NS	NS	15
GRA					NS	NS	14

Coefficients, their standard errors and the statistics of the best model selected are reported. *Ta* Air Temperature, *Precip* Precipitation, *AIK* Akaike's Information Criterion, *NS* not significant

Table 2.3 Results of model selection conducted with the stepwise AIC method using as predictor annual averages of meteorological variables, LAI and CUP

PFT	Rg	Ta	Precip	LAI	CUP	Intercept	R^2adj	*P*	AIC	N
ENF		−17.32 (10.77)		−116.23 (31.06)		276.21	0.64	<0.01	101.8	12
DBF	−79.48 (40.64)			−52.76 (41.37)	−5.39 (1.98)	1,732.9	0.32	<0.05	203.4	14
EBF		95.04 (25.02)	−0.16 (0.10)			−1,783.6	0.66	<0.01	88.99	12
CRO							NS	NS	NS	15
GRA	−28.53 (16.67)	20.19 (8.85)		−64.51 (30.51)	−4.25 (0.37)	824.96	0.95	<0.001	103.12	14

Coefficients, their standard errors and the statistics of the best model selected are reported. *Rg* Shortwave Incoming Radiation, *Ta* Air Temperature, *Precip* Precipitation, *AIK* Akaike's Information Criterion, *NS* not significant

characteristics and phenology in determining the spatial and temporal variability of annual NEE (Table 2.3).

The results of the DStdev computed for each site and grouped for PFT are reported in Fig. 2.6. The time series of DStDev highlights the period that mostly contribute to the inter annual variability of NEE at each site (hereafter referred as the critical period).

As an example, the DStDev computed for IT-MBo site (Fig. 2.5e) shows an abrupt increase in DStDev at the onset of the growing season indicating that this period has an important role in the IAV definition. Marcolla et al. 2011 concluded that climatic conditions and snow cover in late spring largely influence the inter-annual variability of NEE of this alpine grassland, and that the first half of the growing season at IT-MBo mitigates the IAV thanks to the negative correlation ($R = -0.77$) between the timing of snowmelt and the rate of carbon uptake at the beginning of the growing season. A second peak of DStDev due to the variable timing of the meadow cut has been observed, emphasizing the important role of management in controlling the inter annual variability.

The results show that for almost all the PFTs the critical period for the IAV occurs during the spring transient of NEE, when carbon uptake begins after winter, and in some water limited sites at the end of the growing season in late summer.

Table 2.4 Results of model selection conducted with the stepwise AIC method using as predictor annual and sub annual averages of meteorological variables

PFT	Predictors selected	R^2adj	*P*	AIC	AIC (ann. climate)	AIC (ann. climate + LAI and CUP)	N
ENF	*CUP*, **LAI**, *Rg*, *PREC_CUP*, *TA_JJA*, **TA_MAM**	0.98	<0.001	59.73	108.6	101.8	12
DBF	*CUP*, *Rg*, *LAI*, *TA_JJA*, **TA_MAM**, *PREC_JJA*	0.47	<0.05	200.23	NS	203.4	14
EBF	**TA_MAM**, *Rg*, *PREC_CUP*	0.91	<0.001	88.65	88.99	88.99	12
CRO	*VPD_MAM*, *PREC_CUP*	0.44	<0.05	147.85	NS	NS	15
GRA	*CUP*, **TA_MAM**, *Rg*, *LAI*	0.95	<0.001	101.71	NS	103.12	14

PREC Precipitation, *TA* Air Temperature, *Rg* Incoming Shortwave Radiation, *VPD* Vapor Pressure Deficit, LAI and CUP. The averages have been calculated for groups of months (*MAM* March-April-May, *JJA* June-July-August) of for the whole Carbon Uptake Period (e.g. for PREC_CUP). For each PTF the list of selected predictors are reported. In the column 'predictors selected' *bold* variables indicate a positive correlation with NEE positive coefficient) while variables in *italics* indicate negative correlation (negative coefficient). The AIC for the stepwise analysis with annual meteorological variable (Table 2.1) and meteorological variable, LAI and CUP (Table 2.2) are also reported for comparison. *AIK* Akaike's Information Criterion

This has been observed for all the PFTs even if for EBF the signal is weaker and more variable. For CRO the management might be an important confounding factor hampering the identification of the critical period.

Finally, the results of the stepwise AIC analysis is reported in Table 2.4 including as predictor also sub-annual aggregation of climate drivers. These results emphasize that studies focused on the relationship between climate and NEE should investigate the climate variability in particular periods rather than the average climate (Le Maire et al. 2010).

The results showed that for almost all the PFT, the average spring temperature (March, April and May, TA_MAM) is selected as predictor of year-to-year and spatial variations of annual NEE. Cumulative precipitation in summer (June, July and August) or during the entire CUP (PREC_JJA or PREC_CUP) are also selected as predictor of the variability in many PFTs, thus, indicating the water control of NEE in many ecosystems and sites selected.

Moreover, LAI and CUP are also often selected as additional predictors except for EBF and CRO indicating a strong control of LAI, explaining the spatial variability (i.e. across sites) of annual NEE, and phenology, which is one of the main factors controlling the temporal variability of NEE at site level (e.g. Marcolla et al. 2011).

Among the different PFTs, EBF shows the strongest climatic control of the inter annual and spatial variability of NEE. In particular, the variables selected by the stepwise are springtime temperature, precipitation during CUP and radiation. The strong control of climate on EBF has been already reported in a study which showed the effects of hydrological and climatic drivers on a Mediterranean oak forest (Delpierre et al. 2012).

To summarize, the analysis of the relationship between annual NEE and CUP showed that Italian sites sensitivity to CUP variation ranges from -2.97 ± 0.64 to -3.80 ± 0.56 g C m^{-2} yr^{-1} day^{-1}. EBF and GRA are the PFT with the higher portion of inter annual and spatial NEE variability explained by CUP.

Considering the relationship between annual NEE and climate variable, this analysis showed that studies focusing on the relationship between climate and inter annual variations of NEE should concentrate on the variability of climate variables in particular periods of the seasons rather than on the average climate.

2.5 Conclusions

The Eddy Covariance network of Italy provided a great opportunity for a better understanding of the carbon cycle and interaction between ecosystems and atmosphere. In addition, beside the unique collection of measurements that are today available to the benefit of the global scientific community, it has been possible to analyse the relations between NEE and main environmental factors, highlighting the importance of the growing season length and the role of specific climatic variables (mainly temperature and precipitation) in particular periods.

Although the CarboItaly project offering the opportunity to build-up the Italian network has now ended, the teams involved are still trying to do their best in order to continue the measurements collection in the belief that only a long term monitoring activity can provide the basis for an in-depth analysis of how ecosystems will react to climate changes and what their role can be. The importance of these measurements is also confirmed by the construction of the ESFRI (European Strategy Forum on Research Infrastructures) infrastructure ICOS—Integrated Carbon Observing System (www.icos-ri.eu) that will maintain Eddy Covariance sites running across Europe for the next 20 years.

2.6 The Big Effort of Maintaining the Network and Acknowledgments

The maintenance of a network of Eddy Covariance sites like the one described in this chapter has an enormous cost mainly in terms of personnel to be dedicated to the different sites and database. Equipments selection and setup, sensors calibration and control, data collection, data screening, fluxes calculation, database management and additional QA/QC are all fundamental steps that require high level of expertise and preparation. The CarboItaly project provided the broad scientific community with a unique dataset coming from one of the most dense networks of Eddy Covariance sites in Europe; this has been possible thanks to the financial support of the CarboItaly FISR project and a number of European research projects, in particular EuroFlux, CarboEuroFlux, CarboEuropeIP and GHG-Europe. However, all this could not have been possible without the professional and qualified contribution of a number of scientists, PhD students and technicians who devoted large portions of their time to the collection of these measurements, often in difficult meteorological conditions, and then decided to share their data with colleagues around the World in name of the scientific progress. For this reason the main credits of the co-authors of the chapter goes to them, that we want to namely list also to give an indication of the effort needed to collect these measurements: Michael Andersson, Angelo Arca, Nicola Arriga, Narciso Avanzo, Silvia Baronti, Giovanni Callegari, Mario Cammarano, Mauro Cavagna, Alessandro Ciaccia, Roberto Colombo, Ettore D'Andrea, Matteo Danelon, Bruno De Cinti, Giuseppe de Simon, Filippo Di Gennaro, Floriano Di Nardo, Sabina Dore, Dieter Droste, Marlene Duerr, Chiara Ferrè, Alessandro Fiora, Antonio Forgione, Gerardo Fratini, Olga Gavrichkova, Ignacio Goded Ballarin, Gabriele Guidolotti, Francesco Iovino, Mario Lanini, Ramona Magno, Giovanni Manca, Barbara Marcolla, Matteo Mari, Caterina Marino, Serena Marras, Alessandro Matese, Francesco Mazzenga, Franco Meggio, Ana Meijide, Michele Meroni, Stefano Minerbi, Umberto Morra di Cella, Marianna Nardino, Antonia Oriani, Emiliano Pegoraro, Fabio Pieruccetti, Jacopo Primicerio, Nicola Ricca, Emiliano Rosato, Francesco Sabatini, Antonino Sammartano, Giuseppe Santarelli, Giuseppe Scarascia Mugnozza, Andrea Scartazza, Uwe Schwarz, Consolata Siniscalco, Costantino Sirca, Matteo Sottocornola, Luciano Spaccino, Francesco Sposetti, Giampiero Tirone, Michele Tomassucci, Piero Toscano, Andrea Ventura, Andrea Virdiano, Alessandro Zaldei, Roberto Zampedri, Pierpaolo Zara, Terenzio Zenone, Renato Zompanti, Roberto Zorer.

References

Alberti G, Delle Vedove G, Zuliani M, Peressotti A, Castaldi S, Zerbi G (2010) Changes in CO_2 emissions after crop conversion from continuous maize to alfalfa. Agric Ecosyst Environ 136:139–147

Aubinet M, Vesala T, Papale D (2012) Eddy covariance—a practical guide to measurements and data analysis. Springer Science+Business Media B.V. , Netherlands, pp 159–172, ISBN: 978-94-007-2350-4

Beer C, Reichstein M, Tomelleri E, Ciais P, Jung M, Carvalhais N, Rödenbeck C, Arain MA, Baldocchi D, Bonan GB, Bondeau A, Cescatti A, Lasslop G, Lindroth A, Lomas M, Luyssaert S, Margolis H, Oleson KW, Roupsard O, Veenendaal E, Viovy N, Williams C, Woodward FI, Papale D (2010) Terrestrial gross carbon dioxide uptake: global distribution and covariation with climate. Science 329(5993):834–838. doi:10.1126/science.1184984, ISSN: 0036-8075

Ciais P, Reichstein M, Viovy N, Granier A, Ogee J, Allard V, Aubinet M, Buchmann N, Bernhofer C, Carrara A, Chevallier F, De Noblet N, Friend A, Friedlingstein P, Grunwald T, Heinesch B, Keronen P, Knohl A, Krinner G, Loustau D, Manca G, Matteucci G, Miglietta F, Ourcival JM, Papale D, Pilegaard K, Rambal S, Seufert G, Soussana JF, Sanz MJ, Schulze

ED, Vesala T, Valentini R (2005) Europe-wide reduction in primary productivity caused by the heat and drought in 2003. Nature 437(7058):529–533, ISSN: 0028-0836

Chiti T, Papale D, Smith P, Dalmonech D, Matteucci G, Yeluripati J, Rodeghiero M, Valentini R (2010) Predicting changes in soil organic carbon in mediterranean and alpine forests during the Kyoto Protocol commitment periods using the Century model. Soil Use Manag 26(4):475–484. doi:10.1111/j.1475-2743.2010.00300.x, ISSN: 0266-0032

Delpierre N, Soudani K, François C, Le Maire G, Bernhofer C, Kutsch W, Misson L, Rambal S, Vesala T, Dufrêne E (2012) Quantifying the influence of climate and biological drivers on the inter annual variability of carbon exchanges in European forests through process-based modelling. Agric For Meteorol 154–155(15):99–112

FAO, IIASA, ISRIC, ISS-CAS, JRC (2008) Harmonized world soil database (version 1.0). FAO, Rome, Italy and IIASA, Laxenburg, Austria

Gilmanov TG, Soussana JF, Aires L, Allard V, Ammann C, Balzarolo M, Barcza Z, Bernhofer C, Campbell CL, Cernusca A, Cescatti A, Clifton-Brown KJ, Dirks M, Dore S, Eugster W, Fuhrer J, Gimeno C, Gruenwald T, Haszpra L, Hensen A, Ibrom I, Jacobs AFG, Jones MB, Lanigan G, Laurila T, Lohila A, Manca G, Marcolla B, Nagy Z, Pilegaard K, Pinter K, Pio C, Raschi A, Rogiers N, Sanz MJ, Stefani P, Sutton M, Tuba Z, Valentini R, Williams ML, Wohlfahrt G (2007) Partitioning European grassland net ecosystem CO_2 exchange into gross primary productivity and ecosystem respiration using light response function analysis. Agric Ecosyst Environ 121:93–120

Lasslop G, Reichstein M, Papale D, Richardson AD, Arneth A, Barr A, Stoy P, Wohlfahrt G (2010) Separation of net ecosystem exchange into assimilation and respiration using a light response curve approach: critical issues and global evaluation. Glob Change Biol 16:187–208. doi:10.1111/j.1365-2486.2009.02041.x, ISSN: 1354-1013

Le Maire G, Delpierre N, Jung M, Ciais P, Reichstein M, Viovy N, Granier A, Ibrom A, Kolari P, Longdoz B, Moors E.J, Pilegaard K, Rambal S, Richardson AD and Vesala T (2010) Detecting the critical periods that underpin inter annual fluctuations in the carbon balance of European forests. J Geophys Res 115:G00H03

Mahecha MD, Reichstein M, Carvalhais N, Lasslop G, Lange H, Seneviratne SI, Vargas R, Ammann C, Arain MA, Cescatti A, Janssens IA, Migliavacca M, Montagnani L, Richardson AD (2010) Global convergence in the temperature sensitivity of respiration at ecosystem level. Science 329:838–840. doi:10.1126/science.1189587

Marcolla B, Cescatti A, Manca G, Zorer R, Cavagna M, Fiora A et al (2011) Climatic controls and ecosystem responses drive the inter-annual variability of the net ecosystem exchange of an alpine meadow. Agric For Meteorol 151(9):1233–1243

Maselli F, Papale D, Puletti N, Chirici G, Corona P (2009) Combining remote sensing and ancillary data to monitor the gross productivity of water-limited forest ecosystems. Remote Sens Environ 113:657–667. doi:10.1016/j.rse.2008.11.008, ISSN: 0034-4257

Migliavacca M, Reichstein M, Richardson AD, Colombo R, Sutton MA, Lasslop G, Tomellieri E, Wohlfahrt G, Carvalhais N, Cescatti A, Mahecha MD, Montagnani L, Papale D, Zaehle S, Arain A, Arneth A, Black TA, Carrara A, Dore S, Gianelle D, Helfter C, Hollinger D, Kutsch W, Lafleur PM, Nouvellon Y, Rebmann C, Humberto R, Rodeghiero M, Roupsard O, Sebastià M, Seufert G, Soussana JF, Van Der Molen MK (2011) Semi empirical modeling of abiotic and biotic factors controlling ecosystem respiration across eddy covariance sites. Glob Change Biol 17:390–409. doi:10.1111/j.1365-2486.2010.02243.x, ISSN: 1354-1013

Moffat AM, Papale D, Reichstein M, Hollinger DY, Richardson AD, Barr AG, Beckstein C, Braswell BH, Churkina G, Desai AR, Falge E, Gove JH, Heimann M, Hui D, Jarvis AJ, Kattge J, Noormets A, Stauch VJ (2007) Comprehensive comparison of gap-filling techniques for eddy covariance net carbon fluxes. Agric For Meteorol 147:209–232. doi:10.1016/j.agrformet.2007.08.011, ISSN: 0168-1923

Montagnani LG, Manca E, Canepa E, Georgieva M, Acosta C, Feigenwinter D, Janous G, Kerschbaumer A, Lindroth L, Minach S, Minerbi M, Mölder M, Pavelka G, Seufert M, Zeri W Ziegler (2009) A new mass conservation approach to the study of CO_2 advection in an alpine forest. J Geophys Res-Atmos 114:D07306. doi:10.1029/2008JD010650

Niu S, Luo Y, Fei S, Montagnani L, Bohrer G, Janssens IA, Gielen B, Rambal S, Moors E, Matteucci G (2011) Seasonal hysteresis of net ecosystem exchange in response to temperature change: patterns and causes. Glob Change Biol 17:3102–3114. doi:10.1111/j.1365-2486.2011.02459.x

PGAF (2009) Piano di gestione ed assestamento forestale, Piano dei Pascoli e relativa Relazione di Incidenza ai sensi del D.P.R. 357/97 delle proprietà silvo-pastorali dell'Università Agraria di Monte Romano (VT)", Committente Consorzio Forestale Tuscia, Convenzione del 07.05.2007, Approvato dalla Regione Lazio con Determina n. B4110 del 22.09.2009

Papale D, Agarwald DA, Baldocchi D, Cook RB, Fisher JB, Van Ingen C (2012) Database Maintenance, Data Sharing Policy, Collaboration. In: Aubinet M, Vesala T, Papale D (eds) Eddy Covariance—A practical guide to measurement and data analysis. Springer Science+Business Media B.V., Netherlands, ISBN: 978-94-007-2350-4

Papale D, Reichstein M, Aubinet M, Canfora E, Bernhofer C, Longdoz B, Kutsch W, Rambal S, Valentini R, Vesala T, Yakir D (2006) Towards a standardized processing of net ecosystem exchange measured with eddy covariance technique: algorithms and uncertainty estimation. Biogeosciences 3:571–583, ISSN: 1726-4170

Petriccione B, Greco S, Tammaro F (1993) La vegetazione del progettato parco archeologico-naturalistico della Valle di Amplero e della Vallelonga (AQ). Micologia e vegetazione mediterranea VIII:137–160

Reichstein M, Falge E, Baldocchi D, Papale D, Aubinet M, Berbigier P, Bernhofer C, Buchmann N, Gilmanov T, Granier A, Grunwald T, Havrankova K, Ilvesniemi H, Janous D, Knohl A, Laurila T, Lohila A, Loustau D, Matteucci G, Meyers T, Miglietta F, Ourcival JM, Pumpanen J, Rambal S, Rotenberg E, Sanz M, Tenhunen J, Seufert G, Vaccari F, Vesala T, Yakir D, Valentini R (2005) On the separation of net ecosystem exchange into assimilation and ecosystem respiration: review and improved algorithm. Glob Change Biol 11:1424–1439, ISSN: 1354-1013

Reichstein M, Papale D, Valentini R, Aubinet M, BernhoferC, Knohl A, Laurila T, Lindroth A, Moors E, Pilegaard K, Seufert G (2007) Determinants of terrestrial ecosystem carbon balance inferred from European eddy covariance flux sites. Geophys Res lett 34(1), ISSN: 0094-8276

Wohlfahrt G, Anderson-Dunn M, Bahn M, Balzarolo M, Berninger F, Campbell C, Carrara A, Cescatti A, Christensen T, Dore S, Eugster W, Friborg T, Furger M, Gianelle D, Gimeno C, Hargreaves K, Hari P, Haslwanter A, Johansson T, Marcolla B, Milford C, Nagy Z, Nemitz E, Rogiers N, Sanz MJ, Siegwolf RTW, Susiluoto S, Sutton M, Tuba Z, Ugolini F, Valentini R, Zorer R, Cernusca A (2008) Biotic, abiotic and management controls on the net ecosystem CO_2 exchange of European mountain grasslands. Ecosystems 11:1338–1351

Tedeschi et al (2006) Soil respiration in a Mediterranean oak forest at different developmental stages after coppicing. Glob Change Biol 12:110–121. doi:10.1111/j.1365-2486.2005.01081.x

Venables WN, Ripley BD (2002) Modern applied statistics with S. Springer, New York, pp 498. ISBN: 978-0-387-95457-8

Valentini R, De Angelis P, Matteucci G, Monaco R, Dore S, Scarascia Mugnozza G (1996) Seasonal net carbon dioxide exchange of a beech forest with the atmosphere. Glob Change Biol 2:199–207

Yamashita T, Yamashita K, Kamimura R (2007) A stepwise AIC method for variable selection in linear regression. Commun Stat—Theory and Methods 36:2395–2403

Jung M, Reichstein M, Ciais P, Seneviratne SI, Sheffield J, Goulden ML, Gordon Bonan G, Cescatti A, Chen J, De Jeu R, Dolman AJ, Eugster W, Gerten D, Gianelle D, Gobron N, Heinke J, Kimball J, Law BE, Montagnani L, Mu Q, Mueller B, Oleson K, Papale D, Richardson A, Roupsard O, Running S, Tomellieri E, Viovy N, Weber U, Williams C, Wood E, Zaehle S, Zhang K (2010) Recent decline in the global land evapotranspiration trend due to limited moisture supply. Nature 467:951–954. doi:10.1038/nature09396, ISSN: 0028-0836

Wang T, Brender P, Ciais P, Piao S, Mahecha M, Chevallier F, Reichstein M, Ottlé C, Maignan F, Arain A, Bohrer G, Cescatti A, Kiely G, Law BE, Lutz M, Montagnani L, Moors E, Osborne B, Panferov O, Papale D, Vaccari FP (2012) State-dependent errors in a land surface model across biomes inferred from eddy covariance observations on multiple time scales. Ecol Model 246:11–25. doi:http://dx.doi.org/10.1016/j.ecolmodel.2012.07.017, ISSN: 0304-3800

Williams CA, Reichstein M, Buchmann N, Baldocchi D, Beer C, Schwalm C, Wohlfahrt, Hasler N, Bernhofer C, Foken T, Papale D, Schymanski S, Schaefer K. (2012). Climate and vegetation controls on the surface water balance: Synthesis of evapotranspiration measured across a global network of flux towers. Water Resour Res 48(6). doi: 10.1029/2011WR011586, ISSN: 0043-1397

Schulze ED, Wirth C, Heimann M (2000) Managing forests after Kyoto. Science 289:2058–2059

Chapter 3
Biogenic Volatile Organic Compound Emissions

Rita Baraldi, Francesca Rapparini, Osvaldo Facini, Claudia Justina Kemper Pacheco, Giorgio Matteucci, Enzo Brancaleoni and Paolo Ciccioli

Abstract A GIS-based model was developed so to predict Biogenic Volatile Organic Compounds (BVOCs) emissions from the Italian forest ecosystems in order to estimate the fraction of the Net Primary Production lost as reduced carbon and to assess the impact of BVOCs in the formation of ozone and secondary organic aerosols. The performance of the model was verified by comparing the predictions with BVOC fluxes measured in a CarboItaly site using the gradient method with tethered balloon profiles, but also with BVOC fluxes measured in previous years. The agreement between observations and predictions indicated a rather accurate estimation of the model and confirmed the importance in the Italian peninsula of monoterpene emissions, expecially of the fast reacting sabinene in areas dominated by *Fagus sylvatica* L. and *Castanea sativa* L.

R. Baraldi (✉) · F. Rapparini · O. Facini
Institute of Biometeorology, National Research Council (CNR),
Via Gobetti 101, 40129 Bologna, Italy
e-mail: r.baraldi@ibimet.cnr.it

C.J. Kemper Pacheco · E. Brancaleoni · P. Ciccioli
Institute of Chemical Methodologies, CNR,
Via Salaria Km 29.3, C.P 10, 00016 Monterotondo Scalo, RM, Italy

G. Matteucci
Institute for Agriculture and Forestry Systems in the Mediterranean,
CNR, Via Cavour 4-6, 87036 Rende, CS, Italy

G. Matteucci
Institute of Agroenvironmental and Forest Biology, CNR,
Via Salaria km 29, 300 00015 Monterotondo Scalo, RM, Italy

R. Valentini and F. Miglietta (eds.), *The Greenhouse Gas Balance of Italy*,
Environmental Science and Engineering, DOI 10.1007/978-3-642-32424-6_3

3.1 Introduction

Field data indicate that Biogenic Volatile Organic Compounds (BVOC) emitted by vegetation represent a loss of biologically fixed carbon from the terrestrial biosphere (Kesselmeier et al. 2002). Since these losses account for 3–4 % of the Net Ecosystem Production (NEP), they must be taken into account when assessing the amount of CO_2 permanently fixed by forest ecosystems. In areas characterized by high biodiversity, such as the Mediterranean basin, an accurate evaluation of the BVOC emissions can be obtained only if the spatial resolution is high enough to account for the difference in the emission mechanism and emission composition existing between all plant species present in a forest ecosystem (Parra et al. 2004; Simon et al. 2006). High temporal resolution data are also needed because isoprenoids emitted from terrestrial vegetation (mainly isoprene and monoterpenes) react faster than many anthropogenic Volatile Organic Compounds (VOC) with OH radicals and ozone to produce photochemical oxidants (Atkinson and Arey 2003) and secondary organic aerosols (SOA; Arneth et al. 2010; Carslaw et al. 2010). The production of SOA from BVOC seems to significantly affect the earth climate and contribute to its changes (Hoffmann et al. 1997; Odum et al. 1997). In Italy, where conditions leading to photochemical smog episodes extend from the end of March to the end of September (Ciccioli et al. 1989) and levels of ozone and peroxyacetyl nitrate (PAN) downwind large urban areas may reach values as high as 200 and 50 ppbv respectively (Ciccioli et al. 1999), BVOC emission needs to be known at a minimum time resolution of 12 h to model the impact of vegetation on ozone and SOA production. According to Millán-Millán et al. (1996), the extraordinary high ozone levels reached in the Mediterranean basin during the summer are determined by the complex circulation of the air masses occurring under high pressure conditions, when local sea-land breeze wind regimes prevail over large-scale circulation. Under these conditions, the primary emission of NOx and VOC, generated mostly along the Italian coasts during daytime hours, is rapidly transported inland, where forest areas are located. The interaction of anthropogenically polluted air masses with BVOC increases the ozone levels further. The presence of tall mountains (from 1,000 up to 3,000 m a.s.l.) near the coasts (1–80 km) generates a return flow bringing photochemically polluted air masses back to the sea. At night, when advection ceases, ozone stratifies over the sea and is transported again towards the coasts at sunrise, when ozone layers are mixed and the sea breeze is re-activated. In such high reactive conditions, the different reactivity of isoprenoids is crucial for a correct prediction of ozone and SOA production. Emission data expressed in terms of isoprene and total monoterpenes might not always be sufficient for a correct prediction of secondary products. This is particularly true for the Italian peninsula where the emission of monoterpenes is comparable to that of isoprene (Steinbrecher et al. 2009) and composed of compounds displaying rather different potentials for ozone and SOA production (Atkinson and Arey 2003; Hoffmann et al. 1997). The development of a specific isoprenoid emission model is complex due to the different emission mechanisms followed by monoterpenes in Mediterranean plant species. While some deciduous and evergreen trees, such as *Quercus ilex* L. (Ciccioli et al. 1997) and *Fagus silvatica* L. (Dindorf et al. 2006; Moukhtar et al. 2005), emit monoterpenes with the same temperature and light dependent mechanism followed by isoprene in isoprene-emitting plants (Guenther et al. 1995), other evergreen conifers, such as *Pinus cembra* L., emit monoterpenes according to a temperature-dependent mechanism (Steinbrecher et al. 2009). Finally, in some conifers, such as *Pinus pinea* L. (Staudt et al. 2000) and *Abies alba* Mill. (Steinbrecher et al. 2009) both mechanisms appear to be active. Complexity is increased by the fact that the basal emission of isoprenoids (i.e. the emission measured at 1,000 $\mu E\ m^{-2}\ s^{-1}$ of Photosynthetically Active Radiation–PAR–and 30 °C) is not constant through the year, but follows a distinct seasonal trend also in evergreen species (Ciccioli et al. 2003; Staudt et al. 2000; Steinbrecher et al. 2009). To meet both the resolution and reactivity requirements, a GIS-based model was developed in the CarboItaly Project so to predict BVOC emission from the Italian forest ecosystems. The possibility to generate emission maps for isoprene and individual monoterpenes at high spatial (1 km^2) and temporal resolutions (<12 h) was considered to be a fundamental tool to assess the amount of NEP lost in the form of reduced carbon, and to predict the impact of BVOC emission on the production of secondary pollutants in Italy. The performances of the model were verified by comparing the predictions with BVOC fluxes measured in previous years but also with BVOC fluxes measured in a CarboItaly site using the gradient method with tethered balloon.

3.2 Model Description and Its Main Properties

Vegetation maps of Italy with a spatial resolution (grid cell size) of 1 km^2 were produced for the year 2006, by combining the information of the Corine IV Land Cover 2000 (CLC) system with those of the National Forest Inventory (INFC 2005), that was validated through field observations carried out at a very detailed scale (administrative regions). By using this approach, it was possible to define vegetation classes according to their potential for isoprene and individual monoterpene emission. For each vegetation species, a database of isoprene and individual monoterpene emissions was generated by taking into account the different processes leading to their emission. In particular, the relative contribution to the basal emission coming from both the temperature dependent pool (T) and the light and temperature dependent pool (T + L) were assessed for each monoterpene. For the most representative vegetation species present in Italy, such as *Fagus sylvatica, Quercus cerris* L., *Castanea sativa* Mill., *Quercus ilex* L., *Quercus suber* L. and few others, emission values were also verified through laboratory experiments performed at branch levels when the maximum value of the basal emission were reached (July–August). They were performed meeting the requirements given by Niinemets et al. (2011). The basal emission from different pools were determined by measuring the emission at 30 °C in the presence and the absence of the PAR. For the other emitting species, an accurate analysis of the basal emission values reported in the literature was made. Values collected according to the experimental protocol described by Niinemets et al. (2011) were selected when the basal emission reached the maximum value (July–August). This critical analysis showed that, in many cases, the values of the basal emissions used in our model were lower than those used in previous ones (Steinbrecher et al. 2009). Seasonality functions were developed in order to correct the basal emission values for seasonality, taking into due account the sprouting and falling of leaves in Italian ecosystems located at different altitudes. The emission (Ej) of each jth isoprenoid present in the grid cell was calculated using the Eq. 3.1:

$$E_j = \sum_{0}^{n} E_{nj}^{O_{\max} T+L} \times \rho_{nj} \times f_{nj}^{a} \times c_{nj}^{Ta} \times c_{nj}^{L} \times c_{nj}^{S} + \sum_{o}^{n} E_{nj}^{O_{\max} T} \times \rho_{nj} \times f_{nj}^{b} \times c_{nj}^{Tb} \times c_{nj}^{S} \tag{3.1}$$

where n is the number of vegetation species emitting the jth isoprenoid, $E_{nj}^{O_{\max} T+L}$ is the maximum seasonal value of basal emission from the T + L pool, $E_{nj}^{O_{\max} T}$ is the maximum seasonal value of the basal emission from the T pool, ρ_{nj} is the biomass density of the n emitting species, f_{nj}^{a} and f_{nj}^{b} c_{nj}^{Ta} are correction factors accounting for the gradient of light and temperature within the canopy due to sun shading, c_{nj}^{L} and c_{nj}^{Ta} are the empirical coefficients of the Guenther algorithm for plants following a T + L emission mechanism (Guenther et al. 1995), c_{nj}^{Tb} is the empirical coefficient of the Guenther algorithm for plants emitting monoterpenes with a T mechanism, c_{nj}^{S} is the correction term accounting for the seasonal variations of the basal emissions.

In the model, different seasonality terms were adopted for deciduous and evergreen vegetation species. Values of ρ_{nj} in each grid cell were calculated by multiplying weekly values of the LAI, obtained from high resolved satellite observations (MODIS) performed in 2006, for the values of SLA of each one of the vegetation species present in the cell. SLA values were obtained by combining data from the literature with laboratory experiments. Values of on leaf temperature and incoming PAR radiation were collected from satellite observations and elaborated by partners of CarboItaly. Figure 3.1 highlights the high spatial resolution provided by the model, whereas Fig. 3.2 highlights its capability to provide isoprenoid-specific emissions. Both figures were obtained by integrating data elaborated on a 12 h basis. Such a distinction was important because many of the stronger monoterpene emitters followed a T + L mechanism.

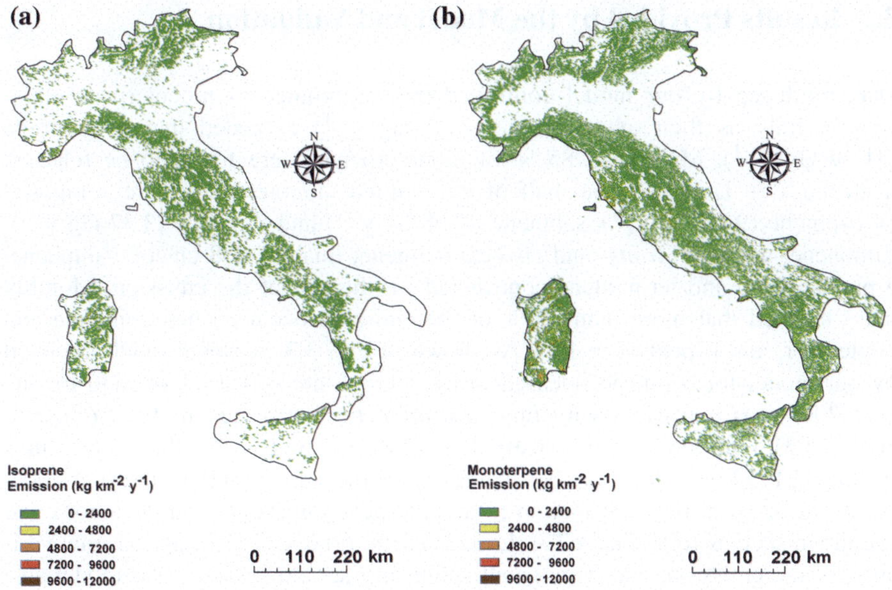

Fig. 3.1 Isoprene (**a**) and monoterpene (**b**) emissions for the year 2006 in Italy

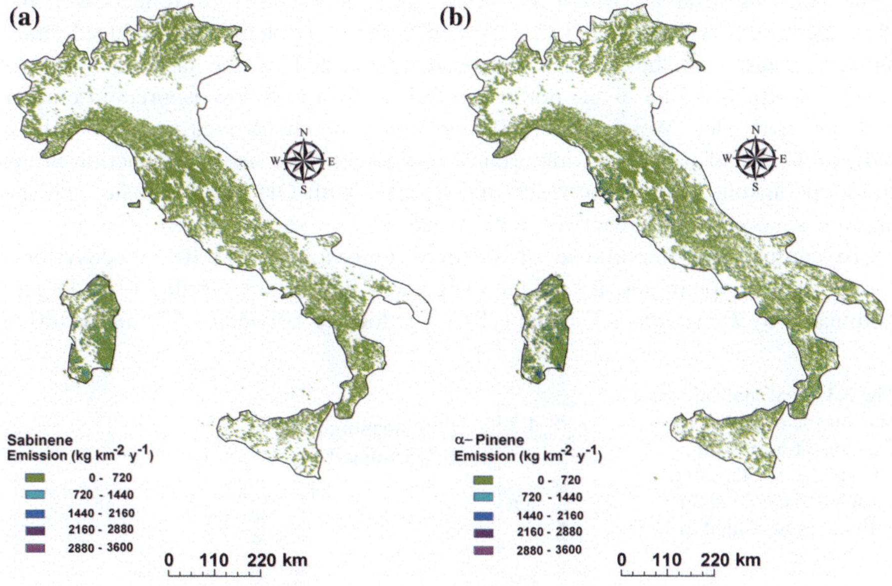

Fig. 3.2 Annual emissions of sabinene (**a**) and of α-pinene (**b**) for the year 2006 in Italy

3.3 Results Provided by the Model and Validation

Data produced by our model confirmed the importance of monoterpene emission in Italy as their annual value (41.2 Gg y^{-1}) exceeded that of isoprene (31.39 Gg y^{-1}). More than 85 % of monoterpenes were found to be released from the T + L pool. About half of the emitted monoterpenes were composed by α-pinene (9.10 Gg y^{-1}), sabinene (4.34 Gg y^{-1}) and β-pinene (3.37 Gg y^{-1}). Limonene, myrcene, *trans*- and *cis*-beta ocimene, linalool, 1-8 cineol, camphene, β-phellandrene and terpinolene contributed for the rest of the emission. Monthly maps showed that more than 70 % of the emission occurred between June and September, and it peaked in July. As shown in Fig. 3.3, seasonal trend predicted by our model for α-pinene released in the coastal site of Castelporziano for the year 2006 was consistent with flux measurements performed in 1997 (Ciccioli et al. 2003). Differences were mostly determined by the fact that the temperatures in August 1997 were much higher than those of the same month of the year 2006. As it can be seen, the contribution to monoterpene emission from deciduous and evergreen species following a T + L mechanism, produced a strong seasonal variation of the emission, with very small values in the cold seasons. Our model also indicates that the largest emission of isoprenoids occurs in Central-Southern Italy and the two Italian islands (Sardinia and Sicily). Isoprenoid emission is concentrated most along the Appennines mountain range separating the east from the west coasts of peninsular Italy. The intense light and temperature emission of sabinene in these areas, which has never been reported before, reflects well the dominance of *F. sylvatica* and *C. sativa* in forest ecosystems. The reduced emission of isoprene in these areas is instead determined by the fact that *Quercus cerris* L., which is one of the dominant species, is a very low isoprenoid emitter (0.1 μg leaf Dry Weight h^{-1}). If confirmed by field measurements, these indications are of paramount importance in assessing the ozone production in the Italian peninsula as sabinene reacts much faster with OH radicals than α-pinene and it has a comparable reactivity with ozone.

To confirm the importance of sabinene emission in the forest ecosystems located in the Apennines, a field campaign was held in a CarboItaly site largely dominated by *F. sylvatica*. The site (IT-Col) is located between 1,500 and 1,700 m

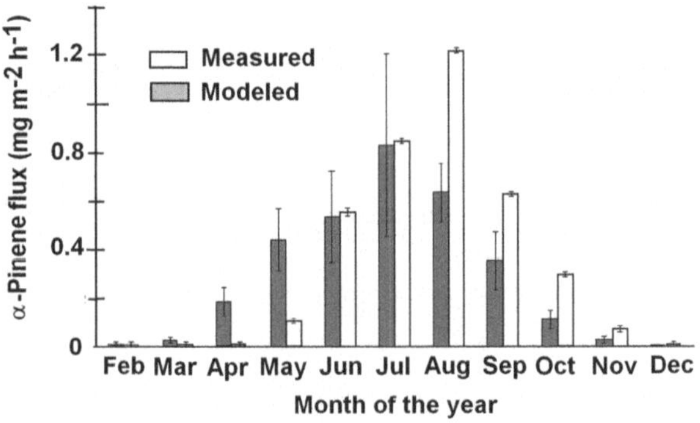

Fig. 3.3 Seasonal variations of α-pinene predicted by the model for the year 2006 in the coastal site of Castelporziano compared with fluxes measured in 1997

a.s.l. between Forchetta Morrea, Prati S. Elia and Selva Piana, near the town of Collelongo (Abruzzo, Italy, see Sect. 2.2.8), at 95 km north-east from Rome. Around the measuring site and at that elevation (1,500–1,700 m a.s.l.) *F. sylvatica* is the dominant tree species for several hundred hectares. The BVOC sampling site, located in a clearing (Prati S. Elia, 700 m wide and 1,500 m long) was 1,000 m a part from the CarboItaly tower in operation since 1993 to measure the ecosystem exchange of water and CO_2 by Eddy Covariance. By considering the complex shape of the area, the gradient method using tethered balloon profiles was the preferred approach to measure VOC fluxes. Only with this approach was indeed possible to collect samples above the average elevation of the canopy and to get values representative of a large area. By collecting VOC at 4 levels from 30 to 200 m above the clearing, sufficient data were available in order to get reliable values of the fluxes. They were measured from July 29th to August 1st 2008. VOC were enriched into cartridges packed with Tenax TA adsorbent resin using automated air sampling systems, equipped with temperature and pressure sensors to maintain a constant flow of air. Vertical profiles of temperature, pressure and relative humidity were also measured. After sampling, the cartridges were taken to the laboratory and analyzed within 10 days by a Gas Chromatography Mass Spectrometry (GC-MS) system equipped with a thermal desorption unit (Baraldi et al. 1999). Using the similarity approach, the Eq. 3.2 was used to calculate VOC fluxes (Kuhn et al. 2007):

$$\text{VOC flux} = \frac{-u \cdot k\left(C_{VOC}^{Z_U} - C_{VOC}^{Z_L}\right)}{\ln\left(\frac{Z_u - d}{Z_L - d}\right) - \Psi_h^{Z_U} + \Psi_h^{Z_L}} \tag{3.2}$$

where Z_U and Z_L are the upper and lower levels (in m) where the VOC concentrations are measured, $C_{VOC}^{Z_U}$ and $C_{VOC}^{Z_L}$ are the concentrations in μg m^{-3} measured at the two levels; K is the Von Karman constant (0.4 in our case); d the average height above canopy where gradient measurements starts; $\Psi_h^{Z_U}$ and $\Psi_h^{Z_L}$ are the adiabatic correction function for heat under unstable conditions, that can be calculated according to Paulson (1970). An example of the vertical profiles collected at the sampling site is shown in Fig. 3.4. Clear gradients were measured for sabinene,

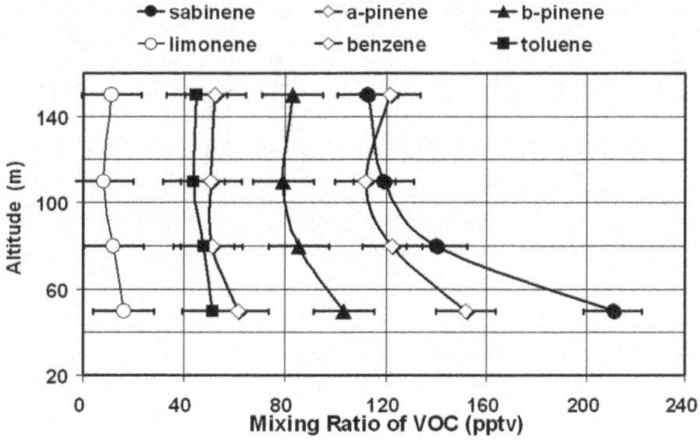

Fig. 3.4 Vertical profiles of 08/01/08 solar time 08:45–09:45 am

α-pinene and β-pinene, and limonene, with the highest concentrations reached by sabinene. An almost homogeneous vertical distribution was observed, instead, for benzene and toluene indicating that no substantial sources for these components existed over the site during these measurements. Data collected on the other days indicated, however, that daily fluxes of α-pinene and β-pinene did not always correlate with those of sabinene, suggesting that other sources than local vegetation contributed to them. In particular night-time trapping of air masses advected from the valleys seemed to simulate well early morning emission. In other cases real emission of these monoterpenes occurred because of biomass burning in picnic fires. In the latter case, high fluxes of α-pinene and β-pinene were accompanied by high fluxes of pyrogenic VOC. Although sabinene is a very photochemically reactive compound (Atkinson and Arey 2003), the turbulent mixing was high enough during sampling to minimize the loss due to reactivity. Severe losses accompanied by strong dilution were indeed observed at an altitude higher than 160 m, where polluted air masses advected from the valleys were mixed with forest emission. By using the concentrations measured with the balloon and the meteorological parameters recorded at the tower, sabinene fluxes were calculated from Eq. 3.2 and compared with the values predicted by the model and with the PAR profile measured at the tower (Fig. 3.5). Data show that measured fluxes of sabinene did not substantially differ from the emission values predicted by the model. Both are fully consistent with the large dominance of *F. sylvatica* over the site and the T + L dependence of its emission (Dindorf et al. 2006). Although both data follow the trend of PAR, some differences exceeding the experimental error are seen in some instances. Since the tower was located 1 km away from the site, we expected such differences to occur whenever meteorological parameters measured at the tower were not exactly the same as those existing over the area where the balloon was raised. This happened when clouds were covering the tower but not the valley. As it can be seen from the PAR profile of Fig. 3.5, the passage of clouds was quite frequent and never was a perfect Gaussian shape centered at noon achieved.

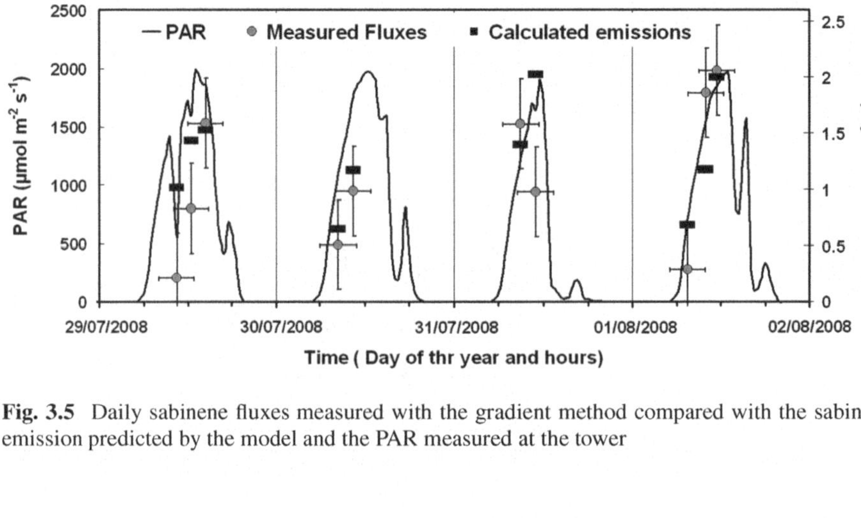

Fig. 3.5 Daily sabinene fluxes measured with the gradient method compared with the sabinene emission predicted by the model and the PAR measured at the tower

Differences between observations and prediction were also expected to arise from the fact that the flat area where the balloon measurements were made was rather small and surrounded by steep slopes and hills. This limitation was impossible to circumvent because there are no flat areas in Italy where strong sabinene emitters grow. Forest areas dominated by *F. sylvatica* and *C. sativa* are, in fact, all located in mountain areas higher than 1,000 m. By considering these limitations, the agreement between observations and predictions can be considered quite good and certainly sufficient to validate the model developed to predict BVOC emissions.

3.4 Conclusions

Table 3.1 summarizes the annual emission rates for different Italian ecosystems that were provided by the model. The validation performed on two CarboItaly sites indicates that the model is rather accurate, suggesting that previous estimates (Steinbrecher et al. 2009) might have overestimated the actual emission. Due to the resolution used and the isoprenoid specific approach followed, the model meets the requirements to assess the impact of BVOC in the formation of ozone and SOA. It is also suitable to assess the fraction of NPP lost as reduced carbon. In this respect, it seems that the carbon emission as BVOC does not account for a substantial fraction of the NPP, as it ranges from 0.03 up to 1.90 %.

Table 3.1 Annual BVOCs emission rates of isoprenoids (Gg y^{-1}) from Italian ecosystems

Vegetation type	Isoprene	Monoterpenes	Total isoprenoids
Forest areas covered by evergreen oaks species (mainly *Q. ilex, Q. suber, Q. coccifera)*	0.06	29.68	29.74
Forest covered by deciduous oaks (mainly *Q. cerris, Q. pubescens, Q. frainetto, Q. petraea, Q. robur*)	29.70	0.09	29.79
Mixed forest areas covered by autochthonous broadleaf species (mainly *Fagus sylvatica, Castanea sativa, Acer platanoides, Fraxinus excelsior, Fraxinus ornus, Carpinus betulus*)	–	7.26	7.26
Forest covered Mediterranean conifers (mainly *P. pinea, P. pinaster, P. halepensis*)	–	1.06	1.06
Mixed forest covered by Mountain conifers (mainly *P. nigra, P. sylvestris, Abies alba, Picea abies, Larix decidua and Pinus cembra*)	1.08	3.20	4.28
Plantation of *Populus sp., Eucaliptus and Olea europea*	0.55	0.01	0.56
Total Italy	31.39	41.29	72.68

References

Arneth A, Harrison SP, Zaehle S, Tsigaridis K, Menon S, Bartlein PJ, Feichter J, Korhola A, Kulmala M, O'Donnell D, Schurgers G, Sorvari S, Vesala T (2010) Terrestrial biogeochemical feedbacks in the climate system. Nat Geosci 3:525–532

Atkinson R, Arey J (2003) Gas-phase tropospheric chemistry of biogenic volatile organic compounds: a review. Atmos Environ 37:S197–S219

Baraldi R, Rapparini F, Rossi F, Latella A, Ciccioli P (1999) Volatile organic compound emissions from flowers of the most occurring and economically important species of fruit trees. Phys Chem Earth Part B 24(6):729–732

Carslaw KS, Boucher O, Spracklen DV, Mann GW, Rae JGL, Woodward S, Kulmala M (2010) A review of natural aerosol formation and feedbacks within the Earth system. Atmos Chem Phys 10:1701–1737

Ciccioli P, Brancaleoni E, Cecinato A, Brachetti A (1989) Diurnal and seasonal variations of Peroxyacetylnitrate (PAN) in a semirural area of Central Italy. In: Brasser G, Mulder WC (eds) Man and his ecosystem. Proceedings of the 8th world clean air congress, Elsevier Science Publ BV, Amsterdam, The Netherlands, vol. 3, pp 497–502

Ciccioli P, Fabozzi C, Brancaleoni E, Cecinato A, Frattoni M, Loreto F, Kesselmeier J, Schafer L, Bode K, Torres L, Fugit JL (1997) Use of the isoprene algorithm for predicting the monoterpene emission from the Mediterranean holm oak *Quercus ilex* L.: Performances and limits of this approach. J Geophys Res 102(D19):23319–23328

Ciccioli P, Brancaleoni E, Frattoni M (1999) Reactive hydrocarbons in the atmosphere at urban and regional scales. In: Hewitt CN (ed) Reactive hydrocarbons in the atmosphere. Academic Press, San Diego, Chapter 5, pp 159–207

Ciccioli P, Brancaleoni E, Frattoni M, Marta S, Brachetti A, Vitullo M, Tirone G, Valentini R (2003) Relaxed eddy accumulation, a new technique for measuring emission and deposition fluxes of volatile organic compounds by capillary gas chromatography and mass spectrometry. J Chromatogr A 985:283–296

Dindorf T, Kuhn U, Ganzeveld L, Schebeske G, Ciccioli P, Holzke C, Köble R, Seufert G, Kesselmeier J (2006) Significant light and temperature dependent monoterpene emissions from European beech (*Fagus sylvatica* L.) and their potential impact on the European volatile organic compound budget. J Geophys Res 111:D16305–1–D16305–15. doi:10.1029/2005JD006751

Guenther A, Hewitt CN, Erickson D, Fall R, Geron C, Graedel T, Harley P, Klinger L, Lerdau M, Mckay WA, Pierce T, Scholes B, Steinbrecher R, Tallamraju R, Taylor J, Zimmerman P (1995) A global-model of natural volatile organic-compound emissions. J Geophys Res—Atmos 100(D5):8873–8892

Hoffmann T, Odum JR, Bowman F, Collins D, Klockow D, Flagan RC, Seinfeld JH (1997) Formation of organic aerosols from the oxidation of biogenic hydrocarbons. J Atmos Chem 26:189–222

INFC (2005) inventario nazionale delle foreste e dei serbatoi forestali di carbonio, estensione e composizione dei boschi (http://www.sian.it/inventarioforestale/jsp/dati_introa.jsp)

Kesselmeier J, Ciccioli P, Kuhn U, Stefani P, Biesenthal T, Rottenberger S, Wolf A, Vitullo M, Valentini R, Nobre A, Kabat P, Andreae MO (2002) Volatile organic compound emissions in relation to plant carbon fixation and the terrestrial carbon budget. Global Biogeochem Cycles 16:1126. doi:10.1029/2001GB001813

Kuhn U, Andreae MO, Ammann C, Araujo AC, Brancaleoni E, Ciccioli P, Dindorf T, Frattoni M, Gatti LV, Ganzeveld L, Kruijt B, Lelieveld J, Lloyd J, Meixner FX, Nobre AD, Poschl U, Spirig C, Stefani P, Thielmann A, Valentini R, Kesselmeier J (2007) Isoprene and monoterpene fluxes from Central Amazonian rainforest inferred from tower-based and airborne measurements, and implications on the atmospheric chemistry and the local carbon budget. Atmos Chem Phys 7:2855–2879

Millán-Millán M, Salvador R, Mantilla E, Artınano B (1996) Meteorology and photochemical air pollution in southern Europe: experimental results from EC research projects. Atmos Environ 30:1909–1924

Moukhtar S, Bessagnat B, Rouil L, Simon V (2005) Monoterpene emissions from Beech (Fagus Sylvatica) in a French forest and impact on secondary pollutants formation at regional scale. Atmos Environ 39:3535–3547

Niinemets U, Kuhn U, Harley PC, Staudt M, Arneth A, Cescatti A, Ciccioli P, Copolovici L, Geron C, Guenther A, Kesselmeier J, Lerdau MT, Monson RK, Penuelas J (2011) Estimations of isoprenoid emission capacity from enclosure studies: measurements, data processing, quality and standardized measurement protocols. Biogeosciences 8:2209–2246

Odum JR, Hoffmann T, Bowman F, Collins D, Flagan RC, Seinfeld JH (1997) Gas/particle partitioning and secondary organic aerosol yields. Environ Sci Technol 30(8):2580–2585

Parra R, Gasso S, Baldasano JM (2004) Estimating the biogenic emissions of non-methane volatile organic compounds from the North Western Mediterranean vegetation of Catalonia Spain. Sci Total Environ 329:241–259

Paulson CA (1970) The mathematical representation of wind speed and temperature profiles in the unstable atmospheric surface layer. J Appl Meteorol 9:857–861

Simon V, Dumergues L, Ponche JL, Torres L (2006) The biogenic volatile organic compounds emission inventory in France. Application to plant ecosystems in the Berre-Marseilles area (France). Sci Total Environ 372:164–182

Staudt M, Bertin N, Frenzel B, Seufert G (2000) Seasonal variation in amount and composition of monoterpenes emitted by young *Pinus Pinea* trees—Implications for emission modelling. J Atmos Chem 35:77–99

Steinbrecher R, Smiatek G, Koble R, Seufert G, Theloke J, Hauff K, Ciccioli P, Vautard R, Curci G (2009) Intra- and inter-annual variability of VOC emissions from natural and seminatural vegetation in Europe and neighbouring countries. Atmos Environ 43:1380–1391

Part II
Forests

Chapter 4
The Role of Managed Forest Ecosystem: An Inventory Approach

Anna Barbati and Piermaria Corona

Abstract The use of statistical sampling coupled with periodic re-measurements in permanent sample units provides the basis for accurate measuring changes in forest extent and conditions, and constructing reliable models to estimate trends. This chapter explores this topic in order to understand how to integrate land use and forest inventory approaches so to estimate CO_2 emissions and removals in managed forest ecosystems. The approach adopted by the Italian National Registry of carbon sinks, targeted to report the activities elected by Italy under the articles 3.3 and 3.4 of the Kyoto Protocol, is based upon the integration of a robust land use inventory, driven by ortho corrected airborne images, and field data from the permanent plots of the national forest inventory. On the whole, the amount of forest land remaining forest in the period 1990–2008 is estimated to exceed 9,000,000 ha; since 1990, over 350,000 ha of croplands and grasslands have been converted to forest, while nearly 127,000 ha have been deforested. The annual change in aboveground living tree biomass in forest land is estimated as high as +14.7 Mt, with an uncertainty (95 % confidence interval) of ±2.8 Mt, which corresponds to an annual carbon gain estimated between 1.35 t C $ha^{-1}y^{-1}$ and 1.45 t C $ha^{-1}y^{-1}$.

4.1 Introduction

Sample surveys are recognized as key tools to estimate changes in land use or in carbon stocks by the Intergovernmental Panel on Climate Change (IPCC) Good Practice Guidance for Land Use, Land Use Change and Forestry

A. Barbati (✉)
Department for Innovation in Biological, Agro-Food and Forest Systems, University of Tuscia, Viterbo, Italy
e-mail: barbati.sisfor@unitus.it

P. Corona
Forestry Research Centre, Consiglio per la ricerca e la sperimentazione in agricoltura, Arezzo, Italy

R. Valentini and F. Miglietta (eds.), *The Greenhouse Gas Balance of Italy*, Environmental Science and Engineering, DOI 10.1007/978-3-642-32424-6_4

(GPG-LULUCF) sector (IPCC 2003). This chapter explores this topic focusing on a specific methodological question: how to integrate land use and forest inventory approaches in the estimate of CO_2 emissions and removals in managed forest ecosystems. The forest carbon accounting system here outlined rests upon two main technical issues arising from the reporting requirements of the Kyoto Protocol for the activities elected by Italy under Article 3.3 (Afforestation/Reforestation and Deforestation—AF/R and D) and 3.4 (Forest Management—FM):

- to provide estimates of the lands that since 1990 are subject to elected LULUCF activities, by geographical units delineated at country level (e.g. administrative units); the system should be able to assign lands to a single activity for each year of the first commitment period 2008–2012;
- to estimate the annual flux of CO_2 to or from the atmosphere on all areas subject to LULUCF activities as identified above; the annual flux of CO_2 is assumed to be equal to annual changes in carbon stocks in biomass and soils carbon pools (IPCC 2003).

The data and information available to a country in order to meet these information needs will largely depend on national circumstances. National Forest Inventories (NFIs) are important data sources, even though not always adequate to meet all the reporting requirements of the Kyoto Protocol. Traditionally, NFIs have been conceived to provide continuously updated information regarding the state of a given country's forest resources, including stock and changes in tree stem volume. But only recently, with increasingly demanding international reporting needs, the question of how NFIs can accurately respond to these new requirements has received specific attention (Corona and Marchetti 2007; Vidal et al. 2008; Tomppo et al. 2010; Corona et al. 2011).

It is *good practice* to use methods providing the highest levels of certainty, while using available resources at country level as efficiently as possible (IPCC 2003). In this perspective, the Italian Ministry of Environment, Land and Sea has chosen an inventory-based approach to set up the National Registry for forest carbon sinks. The Registry is part of the national system for the Italian GHG inventory, which includes all institutional, legal and procedural arrangements for accounting anthropogenic emissions by sources and removals by sinks of GHGs under the United Nation Framework Convention on Climate Change (UNFCCC) and its Kyoto Protocol. As forest management is a key activity for the first commitment period in Italy, the last Italian national forest inventory has been designed to assess the state of carbon stocks in forest land, so that to be named *National Inventory of Forests and forest Carbon pools*—INFC (Gasparini et al. 2010, also refer to http://www.sian.it/inventarioforestale/jsp/home_en.jsp). In consistency with this goal, the INFC adopts a three-phase sampling design for the stratification of the forest land population. The first two phases allow estimating forest area—according to the forest definition selected by Italy under the Kyoto Protocol, i.e. as provided by the Forest Resources Assessment by FAO—and its partition into land use inventory categories and forest types. The third phase provides estimates of carbon stock for all forest carbon pools (living biomass, dead organic matter and soils), scaling up measurements taken on field sample plots of tree biomass correlated variables, deadwood and soil carbon content.

The National Registry for carbon sinks incorporates the INFC set of the first phase sample points within a wider land use inventory system to enable the estimate of GHGs removals/emissions associated to LULUFC activities in Italy and corresponding uncertanities. Inter alia, the Registry is composed of two key components:

- the land use inventory (*Inventario dell'Uso delle Terre*—IUTI) aimed at estimating the area covered by different land use categories, as defined by GPG-LULUCF *good practices*, at given points in time (1990, 2000, 2008, 2012);
- the carbon stock inventory (*Inventario degli Stock di Carbonio*—ISCI) targeted to assess carbon stock changes in forest carbon pools by repeated assessments on the field plots from the INFC third sampling phase.

The following sections are aimed at presenting the methodological framework of the Italian National Registry for Carbon Sinks, as an exemplificative application of the inventory-based approach to forest carbon accounting.

4.2 Building Up a Consistent Time Series by Integration of Land Use and Forest Inventory

The National Registry for Carbon Sinks is conceived to implement a spatial-temporal sampling strategy so to estimate GPG-LULUCF land use categories and the state of forest carbon stocks and their changes over time. The use of formal statistical procedures is a key advantage in comparison to other methods, such as land use mapping; the reliability of the estimates of the quantities of interest can be quantitatively evaluated by the corresponding uncertainties estimates (Corona 2010; Corona et al. 2012; Travaglini et al. 2013). Sampling methods and approaches for the estimation GPG-LULUCF land use classes and changes in forest carbon stock are further discussed in the next sections.

4.2.1 IUTI Land Use Inventory

The amount of lands under FM, AF/R and D activities can fluctuate over time because of various land use changes. Land use transitions are associated with the highest annual changes in carbon stock. Thus, IUTI is targeted to estimate GPG-LULUCF land use categories area at given points in time, from 1990 to the end of the commitment period, while providing adequate levels of statistical accuracy for the administrative units selected for reporting under UNFCC framework in Italy (i.e. district level).

IUTI spatial layout of sampling points follows a tessellation stratified sampling design (also known as unaligned systematic sampling); this sampling scheme is preferable to simple random or systematic grid sampling as sample plots are distributed evenly to all parts of the target area according to a random scheme (Barabesi and Franceschi 2011). The set of sample points is extracted using a 0.5 km square grid, for a total of about 1,206,000 geo-referenced points randomly located in each square cell and fully covering the Italian territory (Fig. 4.1). A subset of the IUTI sample is represented by the 301,300 first phase sample points of the INFC.

IUTI sampling interval and the large sample size are designed to capture and estimate land use changes from 1990 to the end commitment period with a suitable statistical accuracy. Each sample point is photo-interpreted on the basis of digital orthophotos and classified into GPG-LULUCF derived land use classes at given points in time (1990, 2008, 2012). For sample points where land use change in the forest category is detected between 1990 and 2008, as result of AF/R and D activities, the land use classification is performed also at an intermediate point in time (2000), in order to estimate, by interpolation, the annual gain/loss of forest area in different time periods (1990–2000, 2000–2008, 2008–2012). The classification of the sample set is currently completed for the years 1990, 2000 and 2008. The estimation of the area of each land use class at national level, and its breakdown by administrative districts, is performed via area proportions, based on the methodology reported by Corona et al. (2012).

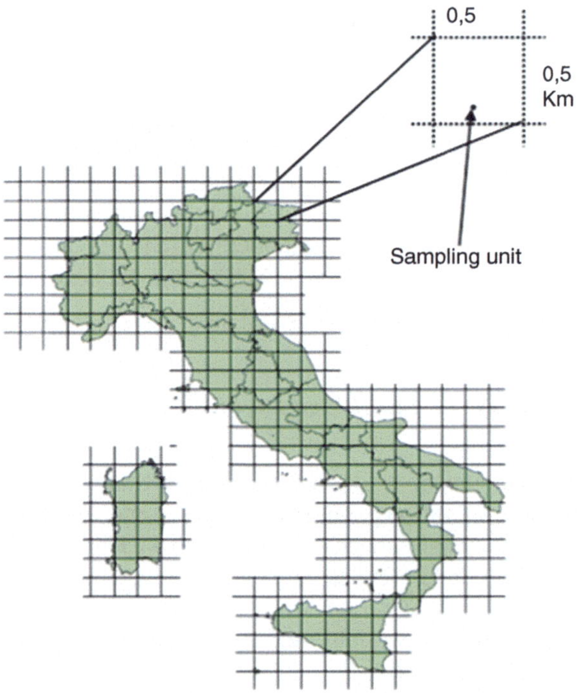

Fig. 4.1 Exemplification of IUTI tessellation stratified sampling scheme; in the background, boundaries of Italy administrative units (district level)

Table 4.1 IUTI land use estimates for the years 1990 and 2008

GPG-LULUCF class	IUTI land use category/subcategory	1990		2008	
		Area (ha)	se %	Area (ha)	se %
Forest land	1. Forest land	9,141,355	0.1	9,653,216	0.1
Cropland	2.1 Arable land	11,315,217	0.1	10,056,141	0.1
	2.2.1 Orchards, vineyards and nurseries	2,682,761	0.3	3,114,765	0.3
	2.2.2 Forest plantations	134,091	1.3	144,376	1.3
Grassland	3.1 Natural grassland and pastures	2,195,754	0.3	1,874,449	0.3
	3.2 Other wooded land	1,867,138	0.3	1,991,200	0.3
Wetlands	4. Wetlands	510,061	0.7	518,586	0.7
Settlements	5. Settlements	1,644,010	0.4	2,140,903	0.3
Other land	6. Other land	658,288	0.6	655,040	0.6

(se % = percent standard error)

Examples of estimates provided by IUTI so far are summarized in Tables 4.1, 4.2 and 4.3. Results show that:

1. the area under different GPG-LULUCF derived land use classes is estimated with high level of statistical accuracy at country level, resulting in a standard error lower than 1 % for most classes (Table 4.1);
2. forest land estimates at district level have high levels of statistical accuracy: the standard error is, by far, lower than 1 %, except for a few administrative units (Table 4.2);
3. as a consequence of the points above, IUTI provides reliable estimates of areas subject to AF/R, D and FM activities (Table 4.3).

On the whole, the amount of forest land remaining forest in the period 1990–2008 is estimated to exceed 9,000,000 ha; since 1990, nearly 220,000 ha of croplands and nearly 400,000 ha of grasslands (including 250,000 of other wooded land) have been converted to forest, while about 127,000 ha have been deforested, for a balance of annual forest expansion equal to 28,000 ha y^{-1}.

4.2.2 *ISCI Carbon Stock Inventory*

ISCI inventory builds upon the INFC third sampling phase by which permanent sample plots are established in the field, centred on points randomly selected from the second phase point sample and stratified by administrative districts and forest types. Field measurements of tree biomass correlated variables (species, tree height and diameter) and deadwood are taken on approximately 7,000 plots, while integrative survey for soil carbon stock evaluation are carried out on a subsample of about 1,500 plots. Field plot survey has been carried in the framework of INFC on 2005 and should be repeated in 2015.

Table 4.2 IUTI forest land estimates at district level for the years 1990, 2000 and 2008

District	1990		2000		2008	
	Area (ha)	se %	Area (ha)	se %	Area (ha)	se %
Abruzzo	412,009	0.6	424,890	0.6	440,267	0.6
Basilicata	312,493	0.7	322,917	0.7	331,667	0.7
Bolzano	357,365	0.6	358,015	0.6	361,115	0.6
Calabria	589,350	0.5	605,270	0.5	606,969	0.5
Campania	458,141	0.6	463,931	0.6	470,995	0.6
Emilia-Romagna	553,084	0.6	574,910	0.6	584,086	0.6
Friuli Venezia Giulia	327,880	0.7	335,476	0.6	337,575	0.6
Lazio	544,532	0.6	556,261	0.5	565,514	0.5
Liguria	368,511	0.5	385,703	0.4	386,253	0.4
Lombardia	601,129	0.6	622,457	0.5	638,216	0.5
Marche	277,506	0.8	305,869	0.7	310,367	0.7
Molise	135,630	1.1	150,739	1.0	154,417	1.0
Piemonte	894,270	0.4	924,895	0.4	939,733	0.4
Puglia	123,576	1.4	138,176	1.3	137,451	1.3
Sardegna	546,851	0.6	574,690	0.6	611,674	0.5
Sicilia	294,836	0.9	308,139	0.8	329,369	0.8
Toscana	1,079,282	0.3	1,116,936	0.3	1,133,810	0.3
Trento	389,612	0.5	393,582	0.5	395,704	0.5
Umbria	363,846	0.6	379,227	0.6	386,480	0.6
Valle d'Aosta	102,102	1.3	105,723	1.3	105,673	1.3
Veneto	409,351	0.7	422,555	0.7	425,881	0.7
Total	9,141,355	0.1	9,470,362	0.1	9,653,216	0.1

(se % = percent standard error)

The approach in the estimate of annual carbon stock change in managed forest land, as quantified by IUTI estimates for each administrative district, is based on the Eq. 4.1 (cf. IPCC 2003):

$$\Delta C_{stockchange} = \frac{C_{t2} - C_{t1}}{t2 - t1} \tag{4.1}$$

where: C_{t1} = carbon stock in the pool at time t1 (tonnes C); C_{t2} = carbon stock in the pool at time t2 (tonnes C).

Annual changes in carbon stocks will be estimated by interpolation of the difference between the state of carbon stocks at the year 2015 (time t2) and the state at the year 2005 (time t1). The estimation of annual changes of carbon stocks in managed land converted to forest land by afforestation and reforestation activities (cf. Table 4.3) requires the assessment of the reference values of carbon stock pools in the previous land use class, like cropland or grassland, besides that of forest carbon stock at the year 2012; this issue will be distinctively approached in Chaps. 10 and 12.

Table 4.3 Land use change matrix 1990–2008, values in hectares (see Table 4.1 for the IUTI codes)

		2008									
		1	2.1	2.2.1	2.2.2	3.1	3.2	4	5	6	Total
1990	1	**9,014,117**	30,192	13,573	975	13,446	37,213	9,497	21,118	1,225	9,141,355
	2.1	***184,398***	9,586,594	789,148	69,470	154,166	128,526	15,374	387,391	150	11,315,217
	2.2.1	***35,547***	272,931	2,269,752	775	21,650	16,571	575	64,962	0	2,682,761
	2.2.2	3,847	51,692	1,249	67,659	2,773	2,349	1,249	3,273	0	134,091
	3.1	***138,121***	60,692	22,573	4,224	1,662,343	276,904	5,349	24,998	550	2,195,754
	3.2	***256,716***	48,566	17,072	750	9,449	1,513,565	7,399	13,097	525	1,867,138
	4	14,696	1,225	425	400	2,999	11,224	476,768	1,500	825	510,061
	5	5,023	4,174	950	125	5,250	3,724	1,250	1,623,439	75	1,644,010
	6	750	75	25	0	2,373	1,125	1,125	1,125	651,691	658,288
	Total	9,653,216	10,056,141	3,114,765	144,376	1,874,449	1,991,200	518,586	2,140,903	655,040	30,148,676

Area of lands subject to FM (*bold*), AF/R (*bold italic*), D (*underlined*) activities as of year 2008 of the commitment period

While waiting for the results from the repetition of the INFC field plot survey in the year 2015, an estimate of the annual change in carbon stock, only for the above ground living biomass carbon pool, has been provided by Tabacchi et al. (2010) on the basis of the so-called default method (IPCC 2003) applied to INFC data taken at the year 2005.

The carbon stock change is calculated as difference between gains and losses in the stem volume of living trees, based on the Eq. 4.2:

$$\Delta C_{default} = I_c + M_{in} + M_{newfor} - M_{harv} - M_{fires} - M_{disturb} - M_{out} \quad (4.2)$$

where: I_c = net annual increment of the stem volume of living trees; M_{in} = stem volume of living trees that at time t_1 are below the minimum dbh threshold, but contribute to a total stem volume at time t_2 (ingrowth); M_{newfor} = stem volume of living trees in areas converted to forest from time t1 to t2; M_{harv} = stem volume of trees harvested from time t1 to t2; M_{fires} = losses of stem volume in forest land due to forest fires; $M_{disturb}$ = losses of stem volume due to natural disturbances; M_{out} = stem volume of living trees in forest land converted to non-forest land from time t1 to t2.

Table 4.4 summarizes the results of the carbon stock change balance in forest lands in Italy for the reference year 2005. Overall, the annual change in above-ground living tree biomass in forest land is estimated as high as +14.7 Mt, with an uncertainty (95 % confidence interval) of ±2.8 Mt.

A comparison with the range of NPP values provided by model-based approaches (see Chap. 5) can be drawn on the basis of the annual carbon gain in the above ground living tree biomass pool as estimated by the net annual increment: this value falls in the 95 % confidence interval from 1.35 t C ha^{-1} y^{-1} to 1.45 t C ha^{-1} y^{-1}; such estimates are both lower than the simulated NPP values for main Italian forest types (i.e. from 3 to 4.5 t C ha^{-1} y^{-1}, see Chap. 5), but such a difference is likely to be explained by the fact that the simulated NPP quantifies carbon stock change for all of the forest carbon pools.

Table 4.4 Estimates and associated 95 % confidence interval (uncertainty) of the annual change of aboveground living tree biomass in the forest land in Italy at the year 2005 (adapted from Tabacchi et al. 2010)

Variable	Estimated value ($Mt\ y^{-1}$)	Uncertainty (%)	Uncertainty ($Mt\ y^{-1}$)
I_c	24.7	±3.7	±0.914
M_{in}	2.4	±30.0	±0.723
M_{newfor}	0.4	±30.0	±0.133
M_{harv}	9.5	±26.2	±2.493
M_{fires}	0.9	±30.0	±0.272
$M_{disturb}$	2.2	±30.0	±0.672
M_{out}	0.2	±30.0	±0.063
Stock change	14.7	±19.4	±2.849

4.3 Conclusions

A current relevant frontier from environmental and economic standpoints is the national accounting to frame policy action toward a greener growth. The purpose of environmental-economic accounting is to assess the sustainability of the economy-ecosystem interaction, not only in terms of conventional indicators like gross domestic product (GDP), but also taking into account the material or energy resource input (and waste generation) to produce one unit of GDP and associated impacts on depletion of natural capital and of related ecosystem services, like carbon sequestration (Corona et al. 2012). Indeed, the improvement of environmental surveys through simultaneous monitoring of multiple attributes is a topic of increasing interest, and it is regarded more favourably by the stakeholders than the establishment and maintenance of separate monitoring and assessment programs. According to these considerations, the approach adopted by the Italian National Registry of carbon sinks is based upon the integration of a robust land use inventory, driven by ortho corrected airborne images (IUTI), and the field data from the permanent plots of the national forest inventory (ISCI). The use of statistical sampling, where sample units are objectively selected by rigorous probabilistic rules as a means of guaranteeing the credibility of estimates, coupled with periodic re-measurements of permanent sample units provides the basis for accurate measuring changes in forest extent and conditions and constructing reliable models to estimate trends; such an approach proves to be an effective tool when estimating changes in forest carbon stocks under the objectives of the United Nation Framework Convention on Climate Change and its Kyoto Protocol.

References

Barabesi L, Franceschi S (2011) Sampling properties of spatial total estimators under tessellation stratified designs. Environmetrics 22:271–278

Corona P (2010) Integration of forest inventory and mapping to support forest management. iForest 3:59–64

Corona P, Marchetti M (2007) Outlining multi-purpose forest inventories to assess the ecosystem approach in forestry. Plant Biosyst 141:243–251

Corona P, Chirici G, McRoberts RE, Winter S, Barbati A (2011) Contribution of large-scale forest inventories to biodiversity assessment and monitoring. For Ecol Manage 262:2061–2069

Corona P, Barbati A, Tomao A, Bertani R, Valentini R, Marchetti M, Fattorini L, Perugini L (2012) Land use inventory as framework for environmental accounting: an application in Italy. iForest 5:204–209

Gasparini P, Tosi V, Di Cosmo L (2010) Development of the Italian national forest unventory. The use and users of the results. Current estimates. Sampling design. Management. Estimation techniques. Options for harmonized reporting. Future prospects. In: Tomppo E, Gschwantner Th, Lawrence M, McRoberts RE (eds) National forest inventories. Pathways for common reporting. Springer, New York , pp 311–331

IPCC (2003) Penman J, Gytarsky M, Hiraishi T, Krug T, Kruger D, Pipatti R, Buendia L, Miwa K, Ngara T, Tanabe K, Wagner F, Good Practice Guidance for Land Use, land-Use Change and Forestry IPCC/IGES, Hayama, Japan

Tabacchi G, De Natale F, Gasparini P (2010) Coerenza ed entità delle statistiche forestali. Stime degli assorbimenti netti di carbonio nelle foreste italiane. Sherwood 165:11–19

Tomppo E, Gschwantner Th, Lawrence M, McRoberts RE (eds) (2010) National forest inventories. pathways for common reporting. Springer, New York

Travaglini D, Fattorini L, Barbati A, Bottalico F, Corona P, Ferretti M, Chirici G (2013) Towards a sampling strategy for the assessment of forest condition at European level: combining country estimates. Environ Monit Assess 185:3255–3268

Vidal C, Lanz A, Tomppo E, Schadauer K, Gschwantner T, di Cosmo L, Robert N (2008) Establishing forest inventory reference definitions for forest and growing stock: a study towards common reporting. Silva Fennica 42:247–266

Chapter 5
The Role of Managed Forest Ecosystems: A Modeling Based Approach

Angelo Nolè, Alessio Collalti, Marco Borghetti, Marta Chiesi, Gherardo Chirici, Federico Magnani, Serena Marras, Fabio Maselli, Costantino Sirca, Donatella Spano and Riccardo Valentini

Abstract Regional approaches to estimate the carbon budget of Italian forest ecosystems using Process-Based Models (PBMs), have been applied by several national institutions and researchers. Gross and net primary productivity (GPP and NPP) have been estimated through the PBMs simulations of carbon, water, and elemental cycles driven by remotely sensed data set and ancillary data. In particular the results of the GPP and NPP estimations provided by the implementation of two hybrid models are presented. The first modeling approach, based on the integration of two widely used models (C-fix and BIOME-BGC), has been applied to simulate monthly GPP and NPP values of all Italian forests for the decade

A. Nolè (✉) · M. Borghetti
School of Agricultural, Forest, Food and Environmental Sciences (SAFE), University of Basilicata, Potenza, Italy
e-mail: nole@unibas.it

A. Collalti · R. Valentini
Impacts on Agriculture, Forest and Natural Ecosystem Division (IAFENT), Euro-Mediterranean Center on Climate Changes (CMCC), Viterbo, Italy

M. Chiesi · F. Maselli
IBIMET-CNR, Sesto Fiorentino, FI, Italy

G. Chirici
Dipartimento di Bioscienze e Territorio (DIBT), University of Molise, Pesche, Italy

F. Magnani
Department of Agricultural Science, University of Bologna, Bologna, Italy

R. Valentini
Department for Innovation in Biological, Agro-Food and Forest Systems (DIBAF), University of Tuscia, Viterbo, Italy

S. Marras · C. Sirca · D. Spano
Impacts on Agriculture, Forest and Natural Ecosystem Division (IAFENT), Euro-Mediterranean Center on Climate Changes (CMCC), Sassari, Italy

S. Marras · C. Sirca · D. Spano
Department of Science for Nature and Environmental Resources (DIPNET), University of Sassari, via E. de Nicola 9, 07100 Sassari, Italy

R. Valentini and F. Miglietta (eds.), *The Greenhouse Gas Balance of Italy*, Environmental Science and Engineering, DOI 10.1007/978-3-642-32424-6_5

1999–2008. The approach, driven by remotely sensed SPOT-VEGETATION ten-day Normalized Difference Vegetation Index (NDVI) images and meteorological data, provided a NPP map of Italian forests reaching maximum values of about 900 g C m^{-2} $year^{-1}$. The second modeling approach is based on the implementation of a modified version of the 3-PG model running on a daily time step to produce daily estimates of GPP and NPP. The model is driven by MODIS remotely sensed vegetation indexes and meteorological data, and parameterized for specific soil and land cover characteristics. Average annual GPP and NPP maps of Italian forests and average annual values for different forest types according to Corine Land Cover 2000 classification are reported.

5.1 Introduction

Simulation models of forest ecosystems answer two needs: first to clarify the relationship between key ecosystem components, for a deeper understanding of their functioning (Kimmins 2008), and second to predict how the state variables of a dynamic system change due to processes in a forest stand or landscape (Brang et al. 2002). In recent years, modeling has undergone significant developments especially in forestry. Modeling tools are increasingly used by both forest ecologists, who face the challenge of transferring knowledge to stakeholders and the general community, and forest managers, who benefit from the development of scenario-based supports for decision-making (Vacchiano et al. 2012). From a general point of view, modeling means trying to capture the essence of a system, deconstructing complex interactions between system components until only the most essential structures and processes remain (Haefner 2005). From stochastic and empirical models, developed over the past 50 years, the increased availability of the data has led to a significant enhancement in the knowledge of the processes that regulate the tree eco-physiology. The difficulties to apply empirical models in sites other than those they were calibrated for, which do not reflect the changes occurred in site conditions or related to management operations since they were developed, have switched to using models able to predict changes in growth and productivity of forests also subject to climate changes, often taking into consideration some factors relating to anthropogenic disturbance. Depending on the modeling purpose, in the last three decades a series of modeling approaches were developed in order to capture forest processes for a wide spatial and temporal resolution scale. The most used approaches are: gap models (Bugmann 2001), landscape models (He 2008), process-based models (PBMs) (Makela et al. 2000) and hybrid models (Zhang et al. 2008). The former of this series explicitly includes site and climate drivers for predicting forest composition, structure and biomass. Small-area or gap models reproduce the growth of single trees within forest patches (e.g., 100 m^2) in relation to the prevailing growth conditions at the site level (Botkin et al. 1972; Shugart 1984; Leemans and Prentice 1989; Pacala et al. 1993). Recent modeling approaches as for the 3D-CMCC

FEM (three Dimensional Forest Ecosystem Model of the euro-Mediterranean Centre for Climate Change) (Collalti et al. 2014) integrates several characteristics of the functional–structural tree models, based on the light use efficiency (LUE) approach, to investigate forest growth patterns and yield processes for complex multi-layer forests.

However, physiological processes are not explicitly accounted for, requiring statistical fitting procedures between each environmental factor and observed growth (Vacchiano et al. 2012).

Landscape models comprise a broad class of spatially explicit models that incorporate heterogeneity in site conditions, neighborhood interactions and feedbacks between different spatial processes (Pretzsch et al. 2008). The aims of these models are to develop scenarios for the sustainability of forest or landscape functions (natural resources, habitat, hydrology, socioeconomic), to forecast their response to disturbances and potential environmental change (climate, N deposition, land use and land use change), to investigate the relationship between landscape structure and regionally distributed risks, and to assess regional-scale matter fluxes, e.g. water, carbon and nutrients. One example is the mesoscale Land Surface Model proposed by Alessandri and Navarra (2008) representing the momentum, heat and water flux at the interface between land-surface and atmosphere; it has been coupled to a general circulation model (GCM) to estimate the rate of forcing by existing vegetation on precipitation patterns. PBMs can be defined as a procedure by which the behavior of a system is derived from a set of functional components and their interactions with each other and the system environment, through physical and mechanistic processes occurring over time (Godfrey 1983; Bossel 1994). More generally, these models are part of the Soil-Vegetation-Atmosphere Transport (SVAT) models giving a representative description of land surface-atmosphere interaction, and describing the physical and biological processes in vegetation and soil, as well as physical processes within the atmospheric boundary layer. SVAT models are commonly used to estimate the exchanges of energy, mass and momentum between the atmosphere and the land surface. These types of models, which are widely applied and validated across the world, use the "big leaf" concept based on one canopy layer or multiple layer schemes, to simulate water and carbon cycles on a variety of spatial (hectare to km) and temporal (daily, monthly or annually) scales. The implementation of these models in forestry in the last decades has been having great success thanks to the availability of remotely sensed data offering a greater amount of information both during the initialization and validation phase. Also, fluxes of energy, CO_2 and water vapor exchanges between the vegetation and the atmosphere measured by the FLUXNET network give the possibility to test such models over as many different circumstances as possible. The spatial scale which they generally work at (ecosystem) can describe the main features in the structure and physiognomy of the forest and they can be considered a valuable tool in the study of those eco-physiological fundamental processes, at species level but also at forest typology level, at an intermediate spatial scale between gap models and Dynamic Global Vegetation Models (DGVM). An important feature of SVAT models is that they can be used as stand-alone models (Marras et al. 2011; Staudt et al. 2011) or as the land surface scheme of a climate model (Pyles et al. 2003). However, as reported by Zhang et al. (2008), most process-based models are unable to simulate forest stand variables (e.g., height, diameter at breast height and volume) since they were not designed for forest management and do not predict forest stand attributes. Battaglia and Sands (1998), Landsberg and Coops (1999) and Makela et al. (2000) have extensively discussed the advantages and disadvantages of using empirical and mechanistic process models. Generally, as postulated by Peng et al. (2002), the weakness of one type of model is the strength of the other, and vice versa. It is almost always possible to find an empirical model providing a better fit for a given set of data due to the constraints imposed by the assumptions of process models. Nevertheless, empirical and process-based models can be combined and integrated into hybrid models in which the shortcomings of both approaches can be overcome to some extent.

According to this general framework, several ecosystem models have been applied to estimate carbon budgets of Italian forests, relying on the availability of remotely sensed data and ancillary dataset provided within the activities of specific national research projects as for the CarboItaly project.

5.2 The Carbon Budget Estimation of the Italian Forests: Ecosystem Models Approach

Regional approaches to estimate the carbon budget of Italian forests have been applied in the last decade by several national institutions and researchers. In particular several PBMs driven by remotely sensed and ancillary data have been applied to run simulations of carbon, water and elemental cycles in order to provide estimates of GPP and NPP and thus of NEP over a wide variety of vegetation types across Italian forest ecosystems.

Over the last decade, the availability of micrometeorological data measured within a national ground-based monitoring network of Eddy Covariance tower sites (flux sites), has been used to calibrate and validate PBMs. In general, the modeling approach is mainly based on the combination and integration of widely applied PBMs into hybrid models so to better represent the high variability of land use, climate and environmental conditions over the Italian territory.

5.2.1 Estimation of Italian Forest NPP. C-Fix and BIOME-BGC Integration Model

Maselli et al. (2009a) and Chiesi et al. (2011) proposed the estimation of forest NPP in Italy based on the integration of a parametric model, C-Fix, and of a bio-geochemical model, BIOME-BGC. C-Fix is a Monteith type parametric model (Veroustraete et al. 2002) which combines satellite-derived estimates of the fraction of Photosynthetically Active Radiation absorbed by forest (fAPAR) with field based estimates of incoming solar radiation and air temperature to simulate total photosynthesis. The annual GPP (g C m^{-2} $year^{-1}$) of a forest can be computed as:

$$GPP = \varepsilon \sum_{i=1}^{12} Tcor_i \cdot Cws_i \cdot fAPAR_i \cdot Rad_i \tag{5.1}$$

where ε is the maximum radiation use efficiency, $Tcor_i$ is a factor accounting for the dependence of photosynthesis on air temperature, Cws_i is the water stress index, $fAPAR_i$ is the fraction of absorbed PAR, and Rad_i is the solar incident PAR, all referred to the i-th month. *f*APAR can be derived from the top of canopy NDVI according to the linear equation proposed by Myneni and Williams (1994). *Cws* was introduced by Maselli et al. (2009a) to optimize the model application in Mediterranean environments, which are characterized by a long and dry summer season when vegetation growth is constrained by water availability. This modification is completed by the use of the MODIS temperature correction factors and the maximum radiation use efficiency equal to 1.2 [(g C MJ^{-1}(APAR))] (Chiesi et al. 2011).

Modified C-Fix was applied to simulate monthly GPP values of all Italian forests for the past decade (1999–2008) following the multi-step methodology described in Maselli et al. (2009a). In summary, a 1-km^2 dataset of monthly minimum and maximum temperatures, precipitation and solar radiation was derived from the available meteorological maps. These maps were further processed to compute the temperature and water stress correction factors which are needed to drive Modified C-Fix. The Spot-VGT ten-day NDVI images of the ten study years were corrected for residual disturbances, composed over monthly periods and processed to obtain *f*APAR maps. All these maps were used to apply Modified C-Fix and yield monthly GPP images over the study years. These images were aggregated to compute an annual average GPP image of Italy, from which average values were extracted for all forest types and Italian Regions.

The ecosystem respirations needed for the prediction of NPP in the Italian forest types were then simulated by BIOME-BGC. This model was developed at the University of Montana to estimate the storage and fluxes of carbon, nitrogen and water within terrestrial ecosystems (Running and Hunt 1993). It requires daily weather data, general information on the environment (i.e. soil, vegetation and site conditions) and on parameters describing the ecophysiological characteristics of vegetation. The model works by searching for a quasi-climax equilibrium (homeostatic condition) with local eco-climatic conditions through the spin-up phase: this means that the sum of simulated respirations become nearly equivalent to GPP, which makes annual NPP approach heterotrophic respiration (Rhet) and NEE tend to zero. Also, such modeling makes the obtained GPP estimates similar to those produced by C-Fix, which are descriptive of all ecosystem components (Maselli et al. 2009b). The version of the model currently used includes complete parameter settings for all main biome types (White et al. 2000). These settings were modified for six forest types to adapt to Mediterranean environments, which show eco-climatic features markedly different from those the model was originally developed for (see Chiesi et al. 2007 for details).

The application of BIOME-BGC in the Italian context required the transformation of the quasi-climax GPP, respiration and allocation estimates into estimates of real forest ecosystems, which are generally far from climax due to the occurred disturbances. The modeling strategy of Maselli et al. (2009b) considers the ratio between actual and potential forest standing volume as an indicator of ecosystem proximity to climax. This ratio can therefore be used to correct the photosynthesis and respiration estimates obtained by the model simulations. Accordingly, actual forest NPP (NPP_A, g C m^{-2} $year^{-1}$) can be approximated as:

$$NPP_A = GPP * FC_A - Rgr * FC_A - Rmn * NV_A \tag{5.2}$$

where GPP, Rgr and Rmn correspond to the GPP, growth and maintenance respiration estimated by BIOME-BGC (g C m^{-2} $year^{-1}$), and the two terms FC_A (actual forest cover) and NV_A (actual normalized standing volume), both dimensionless, are derived from the ratio between actual and potential tree volume.

Due to the previously described functional equivalence of C-Fix and BIOME-BGC GPP estimates, the outputs of the two models can be integrated by multiplying BIOME-BGC photosynthesis and respiration estimates for a ratio between C-Fix and BIOME-BGC GPP. In the current case, BIOME-BGC was applied only to the Tuscany territory, due to the lack of daily meteorological data for the rest of Italy. This required the application of an approximation methodology based on the use of two further assumptions. First, respiration simulated by BIOME-BGC was assumed to vary linearly following photosynthesis, which allowed the calculation of growth and maintenance respiration as constant fractions of GPP for each forest type. Second, a similar assumption was applied to simulate spatial variations of maximum standing volume and LAI, which were needed to compute FC_A and NV_A (Maselli et al. 2009a). Both these assumptions are in reasonable accordance with BIOME-BGC logic, which simulates ecosystems whose all main properties and functions are descriptive of a quasi-climax equilibrium.

The reference values of GPP, respirations, stem carbon and LAI were recovered for each forest type from a BIOME-BGC simulation performed in Tuscany over a 12-year time period (Chiesi et al. 2011). Stem carbon was converted into maximum standing volume using the coefficients given by Federici et al. (2008). BIOME-BGC estimates were then rescaled for each forest type following relevant Modified C-Fix GPP outputs. The regional values of actual forest standing volume needed to compute FC_A and NV_A were extracted for each forest type and Region from the map of Gallaun et al. (2010). All these data were combined within Eq. 5.2 to compute NPP_A for each forest type and Region. CAI values (m^3 ha^{-1} $year^{-1}$) were then computed through Eq. 5.3:

$$CAI = NPP_A * SCA/BEF/BWD * 2 * 100 \tag{5.3}$$

where *SCA* is the Stem C Allocation ratio, *BEF* the volume of above ground biomass/standing volume Biomass Expansion Factor (both dimensionless), and *BWD* is the Basic Wood Density (Mg m^{-3}). The *SCAs* of the six forest types are those of BIOME-BGC, while *BEFs* and *BWDs* are taken again from Federici et al. (2008). The multiplication by 2 accounts for the transformation from carbon to dry matter, and that by 100 for the change in magnitude from g m^{-2} to Mg ha^{-1}.

The CAI modeled values were finally validated through comparison with the CAI measurements taken during the INFC, considering only the Regions where the presence of each forest type was significant (at least 10 1-km^2 pixels). The

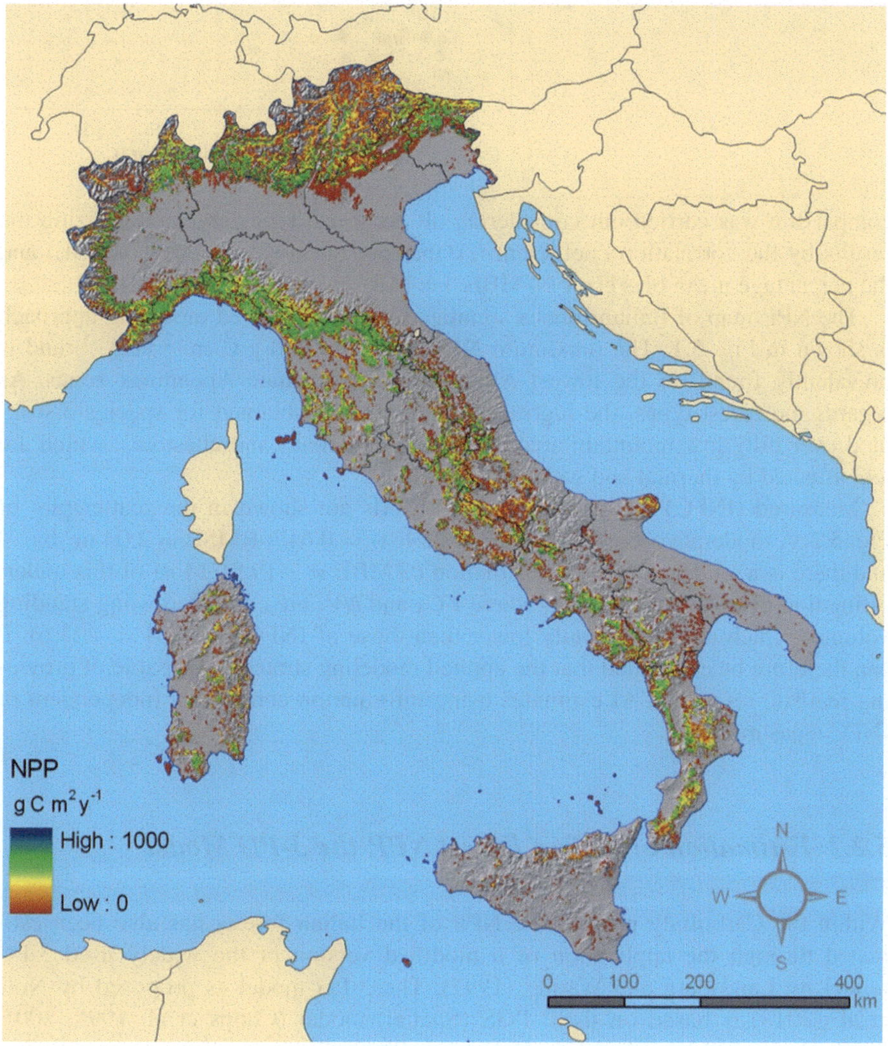

Fig. 5.1 Map of estimated NPP for Italian forests

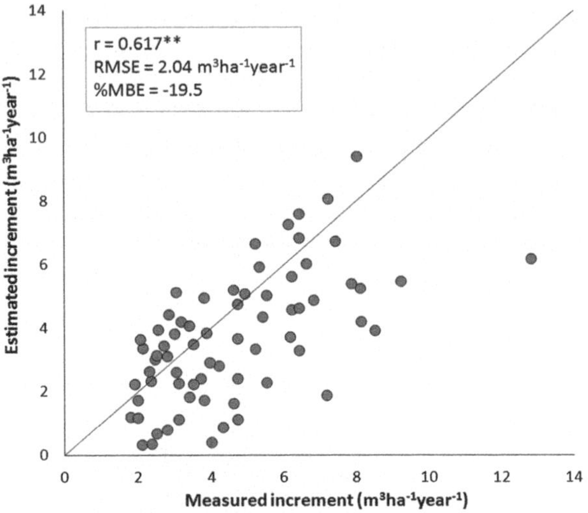

Fig. 5.2 Measured versus estimated forest *CAI* for all forest types and regions considered (n = 69; ** = highly significant correlation, $P < 0.01$)

comparison was carried out considering all six forest types and summarizing the results by the correlation coefficient (r), the root mean square error (RMSE) and the percentage mean bias error (%MBE, i.e. MBE/measured average*100).

The NPP map of Italian forests simulated by the described modeling approach is shown in Fig. 5.1. The maximum NPP is around 900 g C m^{-2} $year^{-1}$, and is prevalently found on the lowest Alpine and intermediate Apennines zones. As regards the forest types, the highest productions are obtained for species distributed over hilly-low mountain areas (i.e. deciduous oaks and chestnut), which are less affected by thermal and water limitations.

Measured (INFC) and estimated forest CAIs are shown in the scatter plot of Fig. 5.2. A moderate accordance is observable ($r = 0.617$; RMSE = 2.04 m^3 ha^{-1}) and there is a tendency to underestimation (%MBE = −19.5). Most of this underestimation derives from Eq. 5.2, where FC_A and NV_A are computed using standing volumes which are significantly lower than those of INFC (%MBE = −23.6). It can therefore be concluded that the applied modeling strategy is capable of providing realistic regional CAI estimates using information completely independent of INFC measurements.

5.2.2 Estimation of Italian Forest NPP. the 3-PG Model

Within the CarboItaly project, the NPP of the Italian forests has also been estimated through the application of a modified version of the widely used 3-PG model by Landsberg and Waring (1997). The 3-PG model as proposed by Nolè et al. (2013) is based on the 3-PGS (Spatial) model (Coops et al. 1998, 2005, 2007; Coops and Waring 2001; Nolè et al. 2009; Tickle et al. 2001) modified to run on a daily time step and produce estimates of GPP and NPP improving model reliability and maintaining the original simplified modeling approach at the same time.

The model fundamental assumption is the canopy LUE (light use efficiency) approach, considering the GPP as the product of the absorbed photosynthetically active radiation (aPAR) and $\varepsilon_{\max}$, which is assumed to be a biome-specific constant for potential LUE (g C m^{-2} MJ^{-1}), and reduced by the effect of environmental constraints ($f_{(x)}$). Daily GPP (g C m^{-2} MJ^{-1}) has then been computed as follows:

$$GPP = aPAR \times \varepsilon_{\max} \times f(x) \tag{5.4}$$

The model reduces daily potential GPP by the effect of environmental constraints represented by four modifiers, ranging between 0 (system "shutdown") and 1 (no constraint). Main environmental modifiers are daily average temperature (T) modifier (f_T), daily VPD modifier (f_D), soil water modifier (f_θ) and light modifier (f_L). Effects of daily average temperature on daily GPP have been modeled as a function of cardinal temperatures, minimum (T_{min}), maximum (T_{max}) and optimum (T_{opt}) temperature for net photosynthetic production, as proposed by Sands and Landsberg (2002). Other environmental modifiers have been calculated according

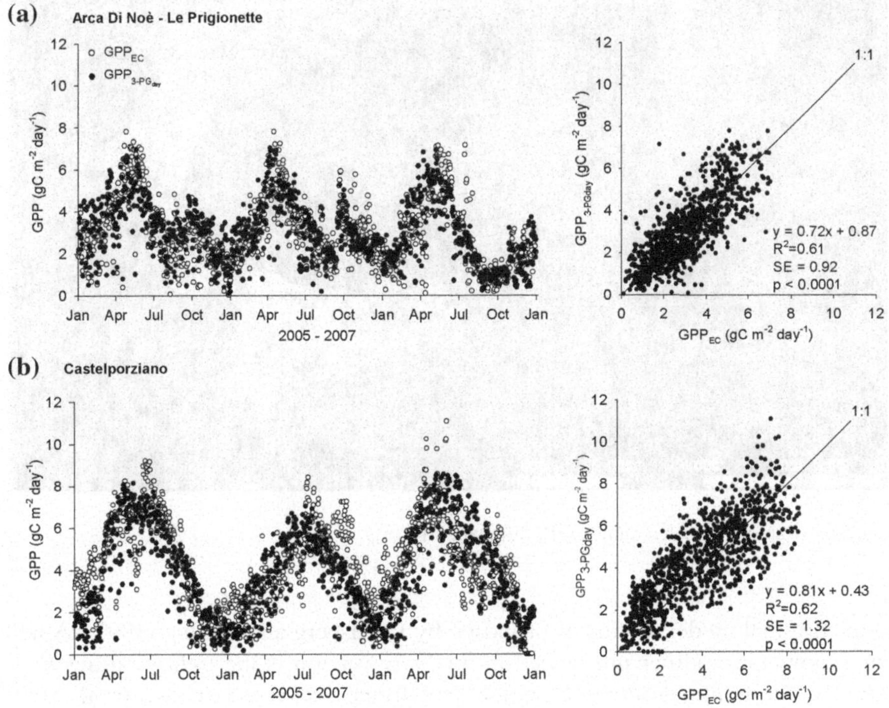

Fig. 5.3 EC measured and 3PG estimated GPP daily patterns for: **a** Arca di Noè–Le Prigionette (2005–2007); **b** Castelporziano (2005–2007) (Nolè et al. 2013)

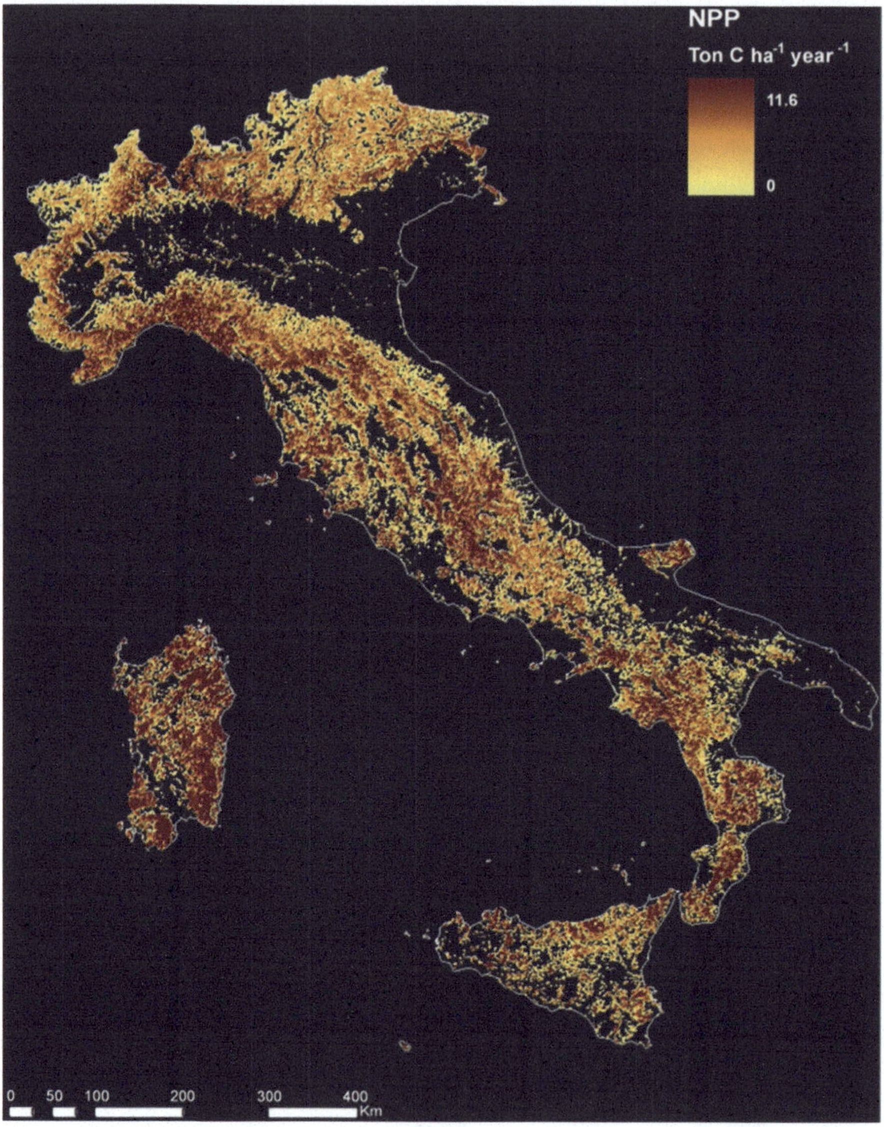

Fig. 5.4 Map of estimated annual GPP for Italian forests

to the original model routine as proposed by Landsberg and Waring (1997). A new environmental modifier introduced in this new version is the light modifier f_L, to describe the nonlinearity light response of forest ecosystem (Grace et al. 1995; Baldocchi and Harley 1995). The light modifier describes, with a hyperbolic function, the gradual saturation of GPP with increasing irradiance, as proposed by Makela et al. (2008):

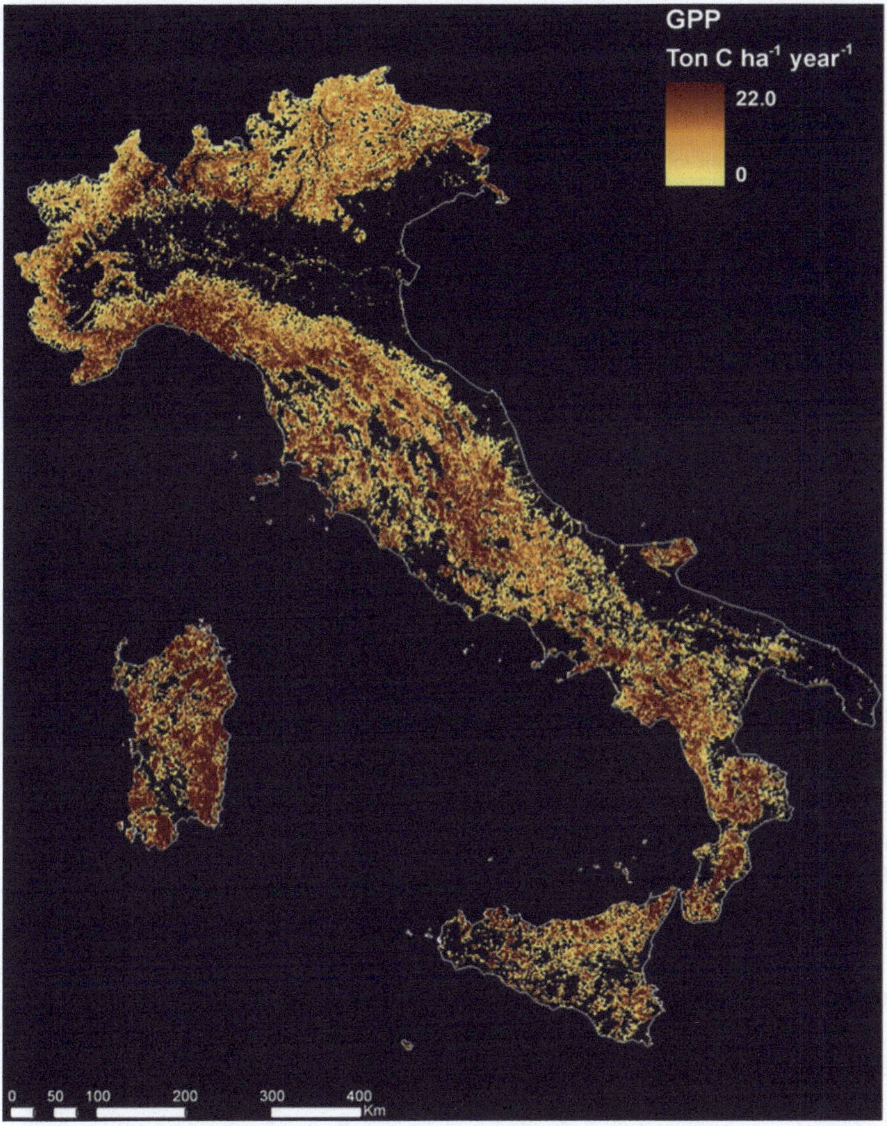

Fig. 5.5 Map of estimated annual NPP for Italian forests

$$f_L = \frac{1}{\gamma aPAR + 1} \tag{5.5}$$

where γ ($m^2 mol^{-1}$) is an empirical parameter.

The input dataset, provided by the partners of the CarboItaly Project, is composed by daily maps of meteorological variables derived from NCEP/NCAR

Table 5.1 Average annual GPP and NPP for different forest types according to Corine Land Cover 2000 classification

Class	Forest type	GPP average (ton C ha^{-1} $year^{-1}$)	NPP average (ton C ha^{-1} $year^{-1}$)
1	Mediterranean shrub land	6.75	3.17
2	Holm oak and evergreen woods	8.93	4.20
3	Woods mainly planted with Mediterranean pine trees and/or cypresses	7.83	3.68
4	Hygrophilous forests	6.81	3.20
5	Broad-leaved woods and plantations with non native species	6.43	3.02
6	Deciduous mixed oaks woods	8.56	4.02
7	Chestnut woods	9.93	4.67
8	Beech forests	8.11	3.81
9	Woods mainly planted with pine-trees in the sub-alpine and alpine areas (silver fir and red fir woods)	8.67	4.07
10	Black pine and mountain pine woods	7.91	3.72
11	Conifers woods and plantations of non native species	8.33	3.92

(Reanalysis) and MSG (Meteosat 2nd generation), remotely sensed vegetation indexes from MOD15A2 LAI–fPAR, soil characteristics from SPADE-2 European Soil Database and land use-land cover maps from Corine Land Cover 2000. Model estimates of GPP have been validated against daily measurements from two Mediterranean Eddy Covariance sites of the CarboItaly Project (Arca di Noè–Le Prigionette (Sardinia) and Castelporziano (Lazio). In particular, model results show a significant correlation (Fig. 5.3) for the Mediterranean sites and a tendency to overestimate GPP during the summer season.

The map of estimated annual GPP and NPP for Italian forests is shown in Figs. 5.4 and 5.5 respectively, with maximum values of forest production mainly distributed in the Apennines sub-alpine areas. In Table 5.1 average annual GPP and NPP for different forest types according to Corine Land Cover 2000 classification is reported, showing the highest values of estimated NPP for chestnut woods, holm oak and evergreen woods, and more generally for low and middle mountain forest ecosystems.

5.3 Conclusions

The implementation of hybrid models, based on the integration of different process-based and empirical models, represents one of the most important tools for the understanding of forest ecosystem processes and to estimate forest ecosystem productivity at regional scale. These models have been applied on a wide range of Italian forest types within several research projects, as for the national specific CarboItaly project.

The availability of both high resolution remotely sensed dataset and micro-meteorological data for model parameterization and validation, contributed to the development of new methodological approaches for the estimation of carbon budgets of Italian forests. The converging results provided by the two different hybrid models previously presented, show the reliability of these models in predicting national forest productivity at regional scale. A significant contribution to models reliability is provided by the availability of ground-based data set measured at the national flux network of Eddy Covariance sites covering main national ecosystem typologies.

References

Alessandri A, Navarra A (2008) On the coupling between vegetation and rainfall inter-annual anomalies: possible contributions to seasonal rain fall predictability over land areas. Geophys Res Lett 35:L02718

Baldocchi DD, Harley PC (1995) Scaling carbon dioxide and water vapour exchange form leaf to canopy in a deciduous forest. II. Model testing and application. Plant Cell Environ 18:1157–1173

Battaglia M, Sands P (1998) Process-based forest productivity models and their applications in forest management. For Ecol Manage 102:13–32

Bossel H (1994) Modeling and simulation. AK Peters Ltd, Wellesley

Botkin D, Janak J, Wallis J (1972) Some ecological consequences of a computer model of forest growth. J Ecol 60:849–873

Brang P, Courbaud B, Fischer A, Kissling-Naf I, Pettenella D, Schoneberger W et al (2002) Developing indicators for the sustainable management of mountain forests using a modelling approach. For Policy Econ 4:113–123

Bugmann H (2001) A review of forest gap models. Clim Change 51:259–305

Chiesi M, Maselli F, Moriondo M, Fibbi L, Bindi M, Running SW (2007) Application of BIOME-BGC to simulate Mediterranean forest processes. Ecol Model 206:179–190

Chiesi M, Fibbi L, Genesio L, Gioli B, Magno R, Maselli F, Moriondo M, Vaccari F (2011) Integration of ground and satellite data to model Mediterranean forest processes. Int J Appl Earth Obs Geoinf 13:504–515

Collalti A, Perugini L, Santini M, Chiti T, Nolè A, Matteucci G, Valentini R (2014) A process-based model to simulate growth in forests with complex structure: evaluation and use of 3D-CMCC Forest Ecosystem Model in a deciduous forest in Central Italy. Ecol Model 272(24):362–378

Coops NC, Waring RH (2001) Estimating forest productivity in the eastern Siskiyou Mountains of southwestern Oregon using a satellite driven process model, 3-PGS. Can J For Res 31:143–154. doi:10.1139/cjfr-31-1-143

Coops NC, Waring RH, Landsberg JJ (1998) Assessing forest productivity in Australia and New Zealand using a physiologically-based model driven with averaged monthly weather data and satellite-derived estimates of canopy photosynthetic capacity. For Ecol Manage 104(1–3):113–127. doi:10.1016/S0378-1127(97)00248-X

Coops NC, Waring RH, Law BE (2005) Assessing the past and future distribution and productivity of ponderosa pine in the Pacific Northwest using a process model, 3-PG. Ecol Model 183(1):107–124. doi:10.1016/j.ecolmodel.2004.08.002

Coops NC, Black TA, Jassal RS, Trofymow JA, Morgenstern K (2007) Comparison of MODIS, eddy covariance determined and physiologically modeled gross primary production (GPP) in a Douglas-fir forest stand. Remote Sens Environ 107:385–401

Federici S, Vitullo M, Tulipano S, De Lauretis R, Seufert G (2008) An approach to estimate carbon stocks change in forest carbon pools under the UNFCCC: the Italian case. iForest 1:86–95

Gallaun H, Zanchi G, Nabuurs GJ, Hengeveld G, Schardt M, Verkerk PJ (2010) EU-wide maps of growing stock and above-ground biomass in forests based on remote sensing and field measurements. For Ecol Manag 260:252–261

Godfrey K (1983) Compartmental models and their applications. Academic Press, New York

Grace J, Lloyd J, McIntyre J, Miranda A, Meir P, Miranda H, Moncrieff J, Massheder J, Wright I, Gash J (1995) Fluxes of carbon dioxide and water vapour over an undisturbed tropical forest in south-west Amazonia. Glob Change Biol 1:1–12

Haefner J (2005) Modeling biological systems: principles and applications. Springer, New York

He (2008) Forest landscape models: definitions, characterization, and classification. For Ecol Manage 254:484–498

Kimmins J (2008) From science to stewardship: harnessing forest ecology in the service of society. For Ecol Manage 256:1625–1635

Landsberg J, Coops N (1999) Modeling forest productivity across large areas and long periods. Nat Res Model 12:383–411

Landsberg JJ, Waring RH (1997) A generalized model of forest productivity using simplified concepts of radiation-use efficiency, carbon balance and partitioning. For Ecol Manage 95(3):209–228. doi:10.1016/S0378-1127(97)00026-1

Leemans R, Prentice I (1989) FORSKA, a general forest succession model. Institute of Ecological Botany, Uppsala

Makela A, Landsberg J, Ek A, Burk T, Ter-Mikaelian M, Agren G et al (2000) Process-based models for forest ecosystem management: current state of the art and challenges for practical implementation. Tree Physiol 20(289):298

Makela A, Pulkkinen M, Kolari P, Lagergren F, Berbigier P, Lindroth A, Loustau D, Nikinmaa E, Vesala T, Hari P (2008) An empirical model of stand GPP with the LUE approach: analysis of eddy covariance data at five contrasting conifer sites in Europe. Global Change Biol 14:92–108

Maselli F, Papale D, Puletti N, Chirici G, Corona P (2009a) Combining remote sensing and ancillary data to monitor the gross productivity of water-limited forest ecosystems. Remote Sens Environ 113:657–667

Maselli F, Chiesi M, Moriondo M, Fibbi L, Bindi M, Running SW (2009b) Modelling the forest carbon budget of a Mediterranean region through the integration of ground and satellite data. Ecol Mod 220:330–342

Marras S, Pyles RD, Sirca C, Paw UKT, Snyder RL, Duce P, Spano D (2011) Evaluation of the advanced Canopy-Atmosphere-Soil Algorithm (ACASA) model performance over Mediterranean maquis ecosystem. Agric For Meteorol 151:730–745

Myneni RB, Williams DL (1994) On the relationship between FAPAR and NDVI. Remote Sens Environ 49:200–211

Nolè A, Collalti A, Magnani F, Duce P, Ferrara A, Mancino G, Marras S, Sirca C, Spano D, Borghetti M (2013) Assessing temporal variation of primary and ecosystem production in two Mediterranean forests using a modified 3-PG model. Ann For Sci 70:729–741. doi:10.1007/s13595-013-0315-7

Nolè A, Ferrara A, Law BE, Magnani F, Matteucci G, Ripullone F, Borghetti M (2009) Application of the 3-PGS model to assess carbon accumulation in forest ecosystems at a regional level. Can J For Res 39(9):1647–1661

Pacala S, Canham C, Silander J (1993) Forest models defined by field measurements: I. The design of a northeastern forest simulator. Can J For Res 23:1980–1988

Peng C, Li J, Dang Q, Apps M, Jiang H (2002) TRIPLEX: a generic hybrid model for predicting forest growth and carbon and nitrogen dynamics. Ecol Model 153:109–130
Pyles RD, Weare BC, Paw U, Gustafso KTW (2003) Coupling between the University of California, Davis, Advanced Canopy–Atmosphere–Soil Algorithm (ACASA) and MM5: Preliminary Results for July 1998 for Western-North America. J Appl Meteorol 42:557–569
Pretzsch H, Grote R, Reineking B, Rotzer T, Seifert S (2008) Models for forest ecosystem management: a European perspective. Ann Bot 101:1065–1087
Running H, Hunt ERJ (1993) Generalization of a ecosystem process model for other biomes, BIOME-BGC, and an application for global-scale models. In: Ehleringer JR, Field CB (eds) Scaling physiological processes: leaf to globe. Academic Press, San Diego, pp 141–158
Sands PJ, Landsberg JJ (2002) Parameterization of 3-PG for plantation grown Eucalyptus globulus. For Ecol Manage 163:273–292
Shugart H (1984) A theory of forest dynamics: the ecological implications of forest succession models. Springer, New York
Staudt K, Serafimovich A, Siebicke L, Pyles RD, Falge E (2011) Vertical structure of evapotranspiration at a forest site (a case study). Agric For Meteorol 151:709–729
Tickle PK, Coops NC, Hafner SD, The Bago Science Team (2001) Assessing Forest Productivity at local scales across a native eucalypt forest using a process model, 3PG-SPATIAL. For Ecol Manage 152(1–3):275–291. doi:10.1016/S0378-1127(00)00609-5
Vacchiano G, Magnani F, Collalti A (2012) Modeling Italian forests: state of the art and future challenges. iForest, e1–e8
Veroustraete F, Sabbe H, Eerens H (2002) Estimation of carbon mass fluxes over Europe using the C-Fix model and Euroflux data. Remote Sens Environ 83:376–399
White MA, Thornton PE, Running SW, Nemani RR (2000) Parameterization and sensitivity analysis of the BIOME-BGC terrestrial ecosystem model: net primary production controls. Earth Interact 4:1–85
Zhang J, Chu Z, Ge Y, Zhou X, Jiang H, Chang J et al (2008) TRIPLEX model testing and application for predicting forest growth and biomass production in the subtropical forest zone of China's Zhejiang Province. Ecol Model 219:264–275

Chapter 6
Emissions from Forest Fires: Methods of Estimation and National Results

Valentina Bacciu, Donatella Spano and Michele Salis

Abstract Emissions from forest fires are recognized to be an important health and environment issue. Fire emissions (FE) include a wide range of gaseous compounds and particles significantly contributing to the atmospheric budgets at local, regional and even global scale. In the last decades, several experimental and modelling studies were carried out to improve knowledge of the atmospheric impact of vegetation fires. FE estimates are affected by several errors and uncertainties; improvements were made possible through new advances in remote sensing, experimental measurements of emission factors and fuel consumption models. In this context, the aim of this chapter is to summarize the state of the research concerning atmospheric FE, highlighting the main methodologies and related uncertainties. In addition, this work presents an overview of historical trends and future scenarios of FE in Italy, starting from the most recent inventories.

6.1 Introduction

Among the primary effects of forest fires, the production of GHGs and solid particulate matter is one of the most important ones (Michel et al. 2005; Ito and Penner 2004). CO_2 and CO, being responsible for about 90–95 % of the total carbon emitted (Andreae and Merlet 2001), are the dominant fractions released by fires. The remaining 5–10 % of carbon emitted is represented by carbonaceous aerosol (35 %), nitrogen oxides (20 %), and CH_4 (6 %) (IPCC 2001). Less than 5 % of the carbon is emitted as particulate matter ($PM_{2.5}$ and PM10, Reid et al. 2005).

V. Bacciu (✉) · D. Spano · M. Salis
Impacts on Agriculture, Forest and Natural Ecosystem Division (IAFENT), Euro-Mediterranean Center on Climate Changes (CMCC), Sassari, Italy
e-mail: valentina.bacciu@cmcc.it

V. Bacciu · D. Spano · M. Salis
Department of Science for Nature and Environmental Resources (DIPNET), University of Sassari, via E. de Nicola 9, 07100 Sassari, Italy

R. Valentini and F. Miglietta (eds.), *The Greenhouse Gas Balance of Italy*, Environmental Science and Engineering, DOI 10.1007/978-3-642-32424-6_6

Fire emissions (FE) interfere with local, regional and global phenomena in the biosphere (Forster et al. 2001; Spichtinger et al. 2001) and influence climate (Randerson et al. 2006; Urbanski et al. 2011). Hydrocarbons and nitrogen oxides can lead to the formation of ozone in smoke plumes, acting as short-lived climate forcer (Urbanski et al. 2011; Naeher et al. 2007). In addition, FE (especially particulates) contributes to air quality degradation and pollution (e.g. Miranda et al. 1994; Schollnberger et al. 2002; Simmonds et al. 2005; Hodzic et al. 2007), sources of significant health and environmental problems. Recent studies have highlighted that fire can be a source of extremely toxic products (e.g. mercury, Friedli et al. 2009a; dioxins EFSA 2012).

The demand for quantification of FE, and the definition of emission inventory, is then increasing due to the need of (1) identification of the source of air pollution affecting human health, (2) accounting for GHG emissions by governments and companies (Bell and Adams 2009), and (3) providing key inputs to air quality and climate change issue analysis and modeling (Granier et al. 2011; Urbanski et al. 2011).

This chapter reviews the contemporary state of the research concerning atmospheric emissions from forest fires. An overview of algorithms and models used to estimate FE, along with the analysis of input uncertainties, is firstly illustrated. Then, a description of emission factors for various GHG gases is proposed. Finally, the analysis of historical trends, temporal and spatial patterns, and future scenarios of FE is presented.

6.2 Quantification of Forest Fire Emissions and Uncertainty Sources

Calculating FE for a specific fuel type (i) requires, according to Seiler and Crutzen (1980), the quantification of the area burned (*BA*, ha), fuel loading (F_l, t ha^{-1}), the proportion of biomass consumed (also known as combustion completeness, CC, %), and the species-specific emission factor (EF). FE is commonly formulated as:

$$FE_i = BA \cdot F_l \cdot CC \cdot EF_l \tag{6.1}$$

The Seiler and Crutzen (1980) approach has been applied, after changes, refinements and integrations, to FE studies from global to local scale and for a variety of aims and purposes (e.g. Battye and Battye 2002; French et al. 2004; Narayan et al. 2007; Wiedinmyer and Neff 2007; Shultz et al. 2008; Wiedinmyer et al. 2010; Thonicke et al. 2010; Carvalho et al. 2011). A common element of these studies and applications is the analysis of uncertainties in the factors of the above mentioned Eq. 6.1. In particular, *BA*, *EF*, and F_l have been pointed out by several Authors as the factors limiting the accuracy of long-term FE data sets (Peterson 1987; Peterson and Sandberg 1988; Battye and Battye 2002; Schultz et al. 2008). To represent their variability then becomes the challenge in FE inventories development (Langmann et al. 2009).

Large systematic errors in BA assessment may exist in relation to the reporting system (Peterson 1987). Barbosa et al. (1999), Korontzi et al. (2004), Al-saadi et al. (2008) analyzed the burnt area uncertainty and its effects of FE using different spatial datasets. New advances in remote sensing have made datasets of active fires and burned areas widely available, and have also provided temporal and spatial resolution improvements (Ottmar et al. 2009), but the presence of clouds or confused signals could strongly affect data (Stroppiana et al. 2010); Friedli et al. (2009b).

Fuel load (F_l) is considered as the source of the largest errors in FE estimates (Peterson and Sandberg 1988; Peterson 1987; Hardy et al. 2001), due to the large variability in ecosystem types and species composition. For example, Dimitrakopoulos (2002) pointed out the high variability of fuel load within Mediterranean vegetation, ranging from 4.85 t ha^{-1} in grasslands to 53 t ha^{-1} in evergreen sclerophyllous shrublands (fuel depth: 1.5–3 m). Recently, to obtain a better simulation of spatial variability in fuel load, several Authors integrated biochemical models (van der Werf et al. 2006) or dynamic global vegetation models (Thonicke et al. 2010) in FE inventories.

Peterson (1987) and Peterson and Sandberg (1988) demonstrated that EF variability (mainly due to type of pollutant, type and arrangement of fuel, and combustion efficiency, CE), contributes to about 16 % of the total error associated with emissions. The uncertainty is higher for compounds and biomes that have not been studied in detail (Langmann et al. 2009). Ward and Hardy (1991) suggested that EF could be better assessed considering the two phases of combustion, and then the fraction of biomass consumed by each process. The CE concept was then introduced, in order to involve the fractional rate of complete combustion. Values of CE exceeding 90 % indicate the flaming phase, while values lower than 85 % indicate smoldering combustion (Ward and Hardy 1991; Yokelson et al. 2007).

Another key component in the estimate of the amount and source of emissions (Ottmar et al. 2009) is the combustion completeness (CC, in %) (Shea et al. 1996). CC represents the ratio between consumed and available fuel load, and depends on fire type, fuel type and its moisture content (Ward et al. 1996; Battye and Battye 2002; Langmann et al. 2009). CC of coarse and wet fuels is lower than fine and dry fuels. Most FE estimation models determine combustion factors for each fuel strata (e.g. stems, leaves and litter) and moisture conditions, allowing for a detailed description of CC (e.g., Reinhardt et al. 1997; Hardy et al. 2001; van der Werf et al. 2006). In order to increase the level of accuracy for the assessment of CC, Chiriacò et al. (2013) integrated a methodology proposed by Bovio (2007) based on the level of damage assessed on the basis of forest vegetation class and scorch height, which depends on two main factors: the intensity of the fire and the type of forest vegetation affected by fire.

6.3 Description of Emission Factors for GHGs

According to Andreae and Merlet (2001), the mass of pollutant produced (M_x, in g) per mass of dry fuel consumed (M_b, in kg) is referred to as emission factor of a chemical species x (EF_x) (g kg^{-1}), which translates biomass burned into trace species emissions (see Eq. 6.2).

$$EF_x = \frac{M_x}{M_b} \tag{6.2}$$

Since the '80s, a number of large projects dealing with biomass burning experiments and EFs have been coordinated in different ecosystems (Andreae and Merlet 2001). Most of these integrated projects focused on rainforest and savannas in Africa (e.g. Lacaux et al. 1995; Hao et al. 1996; Delmas et al. 1999; Swap et al. 2003) and Brazil (e.g. Crutzen et al. 1985; Kaufman et al. 1998; Yokelson et al. 2007). The mentioned field campaigns provided highly scattered information through the literature (Andreae and Merlet 2001); due to the different adopted methods of EF measurements, these data also present a large variability. In 2001, Andreae and Merlet reviewed and summarised EFs for a broad number of chemical species and for various types of biomass burning, such as savanna and grassland, tropical forest, extra-tropical forest.

Except for very few studies, there is a lack of available emission data from boreal zone and temperate forest, despite the importance of fire in these areas (Koppmann et al. 2005; Urbanski et al. 2009). Recently, Urbansky et al. (2009) compiled EF data for five broad vegetation cover types, considering temperate forest too. EF data from this ecosystem type mostly derived from experimental and prescribed fires in the United States and Canada, typically ignited under favorable weather and fuel conditions in order to manage low intensity fires.

In Mediterranean areas, Miranda et al. (2005b) collected data of emission from shrubland experimental fires in the framework of the SPREAD project (Forest Fire Spread Prevention and Mitigation 2002–2005 founded by EGVI-2001-00027). More recently, Alves et al. (2010, 2011) detailed particle and trace gas EF for prescribed fires in Mediterranean shrubland and wildfires of the 2009 summer season, respectively. Fig. 6.1 compares several shrubland EFs for $PM_{2.5}$, CO, CO_2, CH_4.

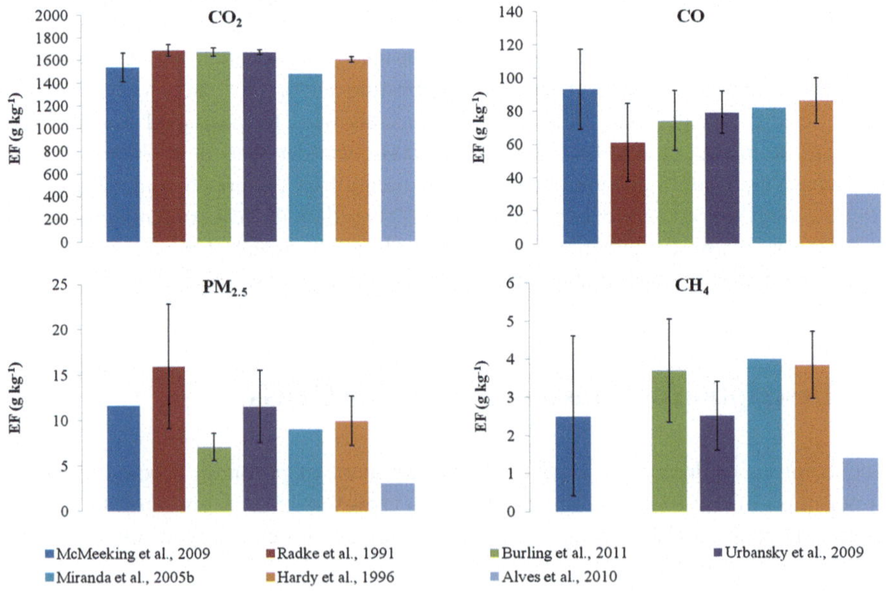

Fig. 6.1 Comparison among shrubland emission factors (g kg^{-1}). Data derived from literature

6.4 Fire Emissions Modelling

According to Debano et al. (1998), quantitative predictions of fire effect models can be distinguished based on the fuel consumption modelling approach. Empirical models (McRae 1980; Brown et al. 1991; Prichard et al. 2005) use statistical relationships derived from measured woody fuel consumption data (Hollis et al. 2010) while physical models fully describe heat transfer processes (Albini 1976a). Semi-physical models result in the combination of these two approaches (Albini et al. 1995; Albini and Reinhardt 1995, 1997).

Up to now, among a conspicuous number of fuel consumption models, CONSUME (Fuel Consumption model, Ottmar et al. 1993, 2006) and FOFEM (First Order Fire Effect Model, Reinhardt et al. 1997) are those mostly used. CONSUME falls into the empirical model category, predicting fuel consumption by combustion phase, heat release and pollutant emissions (Prichard et al. 2005). FOFEM employs the physical model of heat transfer BURNUP (Woody fuel consumption model, Albini et al. 1995; Albini and Reinhardt 1995) to predict woody and litter fuel consumption and heat release.

At regional scale, Urbanski et al. (2011) presented the 2003–2008 Wildland Fire Emission Inventory (WFEI) in the contiguous United States. The product combines observation from satellite data, fuel loading maps, fuel consumption models (both COMSUME and FOFEM), and an EF database. Air pollutant emissions associated with forest burning in Texas were estimated by Dennis et al. (2002), based on survey and field data on area burned and land covered, and FOFEM for fuel consumption and emission factors.

At local scale, several case studies have been performed to determine the impact of fire events on air quality and identify the source of air pollution. Clinton et al. (2006) implemented FOFEM algorithms to quantify the source and composition of smoke and emissions from wildland fires that affected Southern California, USA, in October 2003. French et al. (2011) applied the two models to compare the different methodologies of FOFEM and CONSUME to estimate carbon loss from terrestrial biosphere resulting from wildland fires in Canada. Bacciu et al. (2012) applied FOFEM in Mediterranean areas, estimating type and amount of Mediterranean vegetation fire emissions from Sardinian fires (2005–2009), and compared 2005 emission estimates with the Italian Emission Inventory (NEI-PROV, De Lauretis et al. 2009) (Table 6.1).

Table 6.1 Average pollutant and GHG mass (Gg) by category from Sardinian fires (2005–2009)

YEAR	PM_{10}	$PM_{2.5}$	CH_4	CO	CO_2	NO_X	SO_2	TOTAL
Mean	0.78	0.66	0.32	5.97	200.97	0.32	0.12	209.15
2005	0.41	0.34	0.16	2.98		0.18	0.07	4.14
NEI-PROV	0.62	0.62	0.34	3	–	0.09	0.03	4.74

The total emissions estimated for the 2005 fire season were compared with the 2005 fire emissions from NEI-PROV (modified from Bacciu et al. 2012). (CO_2 emissions from fires not present in the NEI-PROV report)

6.5 Fire Emission Historical Trends and Future Scenarios

A crucial point to support emissions inventories and modeling of air quality and climate change issues has been recognized in up-to-date, accurate, and consistent FE estimates (Battye and Battye 2002; Granier et al. 2011). In addition, the spatial distribution pattern quantification is decisive for several applications, including air quality management plans, and emission source models coupled with dispersion models and decision support systems (Bacciu et al. 2012).

Over the past decades, a number of inventories, both at global and regional scale, were developed to estimate gaseous and particulate species emissions from forest fires. Earlier studies (e.g. Crutzen and Andreae 1990; Hao et al. 1990; Galanter et al. 2000) made use of biome-averaged fuel load and fire return times, while the latest inventories used satellite remote-sensing data to derive burned area or active fires, often combined with biogeochemical models (e.g. Hoelzemann et al. 2004; Ito and Penner 2004; van der Werf et al. 2006, 2010).

In this section, some new inventories (Tables 6.2 and 6.3) were analyzed with the aim to give an overview of the historical trends, as well as future scenarios of FE in Italy.

Table 6.2 Long-period FE inventories description

Inventory	*RETRO*	*EDGAR4.2*
Time coverage	1960–2000	1970–2008
Time resolution	Monthly	Yearly
Grid size	0.5°	0.5°
Fire	National statistics; GBA2000; ATSR fire pixels	(1970–1996) RETRO; Fire counts (1997–2000) ATSR; Fire counts (2001 onwards) MODIS; Burned area (2001 onwards) MODIS
Reference	Shultz et al. (2008)	EC 2011

Table 6.3 Monthly based fire emission inventories description

Inventory	Time coverage	Time resolution	Grid size	Fire product	Reference
GFED3	1997–2010	M	0.5°	ATSR, VIRS, and MODIS	van der Werf et al. (2010)
GICC	1900–2005	D/M	1.0°	GBA2000, ATSR, historical reconstruction of BA from Mouillot and Field (2005)	Mieville et al. (2010)
GFED2	1997–2005	M	1.0°	ATSR, VIRS, and MODIS	van der Werf et al. (2006)
GUESS-ES	1997–2009	M	1.0°	GFED3 Burned area, L3JRC from SPOT-VEGEATION	Knorr et al. (2011)

In the first part, past trends of CO_2 and CO from fires were evaluated from two long-period inventories (data supplied by www.eccad.sedoo.fr), overlapping for the period 1970–2000. CO and CO_2 were selected due to their role as greenhouse gasses, ingredients of smog chemistry, and their contribution on total carbon emitted by fires.

Then, we illustrated the differences and uncertainties in emission amount and spatial extend of emission estimates from five inventories at monthly basis, compared with the burned area pattern as recorded at provincial level by JRC.

Finally, 21st century FEs for four species were illustrated, starting from the Representative Concentration Pathways (RCPs). The RCPs are a set of four new possible pathways of emissions and land use developments, based on consistent scenarios representative of existing literature (van Vuuren et al. 2011).

6.5.1 CO and CO_2 Historical Trends

The basic characteristics of the biomass burning inventories used in this work are summarized in Table 6.2, and other information can be found at www.eccad.sedoo.fr.

The RETRO inventory (Shultz et al. 2008) is based on the analysis of available literature and datasets, estimates from different satellite products, and a semi-physical numerical model to simulate fire occurrence and spread. EDGAR4.2 (EC 2011) estimates FE from large scale biomass burning (savanna, grassland, and forest fires). It was developed on GFED2 inventory (van der Werf et al. 2006) for the period 1997–2005, while the data before 1997 were derived from GFED2 inventory scaled back with regional biomass burning trend from RETRO.

As for the temporal pattern, the two datasets seemed to well capture the inter-annual variability showed by the Italian burned areas (BA) from 1970 to 2000 (Fig. 6.2); peaks on FE estimations corresponded to peaks in burned area. For example, the RETRO maximum value (2.9 Tg CO_2 and 0.18 Tg CO) was in 1990, when $1.95*10^5$ hectares burned. Nonetheless, the extrapolation of a clear and reliable trend would be hardly possible. RETRO inventory showed a monotonic trend, characterized by a smooth inter-annual variability (range of 2.04 Tg CO_2 and variation coefficient of 0.40 %), while the variability was more marked in EDGAR4.2 (range of 6.48 Tg CO_2 and variation coefficient 0.37 %). The differences in CO estimations were smaller than CO_2; in 1970 the difference was about 55 % against 73 % for CO and CO_2, respectively. In 1990, the difference between the two products was still high for CO_2 (66 %) and smaller for CO (44 %), while in 2000 it was about 56 and 23 %, respectively.

6.5.2 Fire Emission Amount and Spatial Distribution

With the aim to emphasize differences in the amount and spatial distribution of FE estimates from five inventories at monthly basis, maps of CO emissions for July 2000 are presented in Fig. 6.3. The inventories used in this comparison, obtained

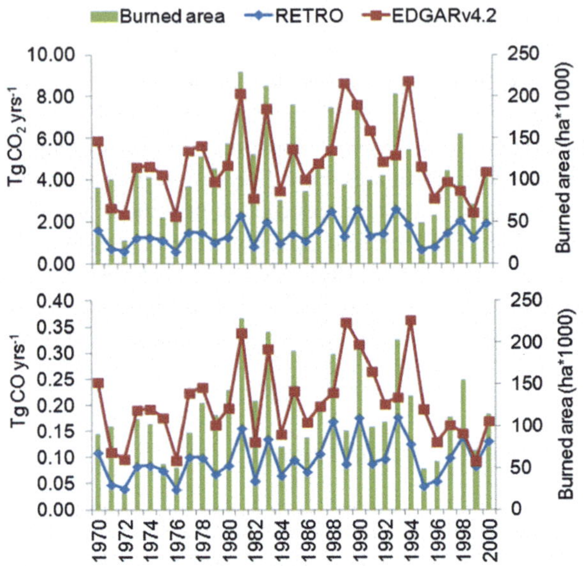

Fig. 6.2 Comparison among annual fire emissions from 1970 to 2000 of CO_2 (*top panel*) and CO (*bottom panel*) from RETRO and EDGARv4.2 inventories

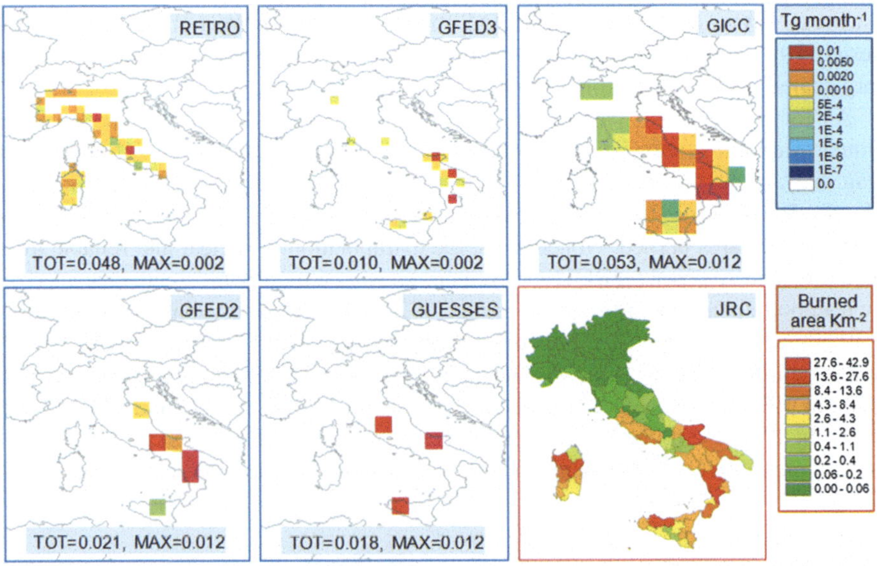

Fig. 6.3 Maps of CO emissions (Tg month^{-1}) from five inventories at monthly base (RETRO, GFEDv3, GICC, GFEDv2, GUESS-ES) and NUT03 burned area (km^2) for July 2000. RETRO and GFEDv3 are at 0.5° resolution, GICC, GFEDv2, GUESS-ES are at 1° resolution

from the website www.eccad.sedoo.fr, are RETRO, GFEDv3 (van der Werf et al. 2010), GICC (Mieville et al. 2010), GFEDv2 (van der Werf et al. 2006), GUESS-ES (Knorr et al. 2011) (see Tables 6.2 and 6.3 for basic characteristics). We also compared spatial distribution from the five EF inventories with the burned area pattern (km^2) as recorded at provincial level by JRC.

The largest monthly burned area was observed in Foggia (Apulia region, 43 km^2), Catanzaro (Calabria region, 39 km^2), Cosenza (Calabria region, 28 km^2), Palermo (Sicily, 22 km^2), and Nuoro (Sardinia region, 17 km^2) (Fig. 6.3). GFEDv3, GICC, GFEDv2, and GUESS-ES agreed, on the whole, on the spatial distribution of the biomass burning emission (as shown by Fig. 6.3) and on the amount of CO emitted, despite the different grid size.

For example, as regards the Foggia burned area, GUESS-ES estimated the highest value of biomass burning, from 0.005 to 0.01 Tg CO, followed by GFEDv3 (from 0.002 to 0.005 Tg CO). GICC and GFEDv2 assessed the lowest values, from 0.001 to 0.002 Tg CO. Furthermore, as per the Catanzaro burned area, GFEDv3, GICC, and GFEDv2, unlike RETRO database, agreed on spatial location and the total amount of CO emitted (0.002, 0.012, and 0.012 Tg CO respectively). Finally, only RETRO inventory recorded biomass burning due to large fires in Sardinia Island.

Differences among emission inventories are likely because of different products to estimate burned area, as highlighted by Jain (2007) at both regional and global scale. For example, in Italy RETRO inventory uses long-term annual fire statistics at province level, and the geographic distribution derive from qualitative literature description and random distribution of individual large fires (Shultz et al. 2008). Nevertheless, the RETRO Authors admitted that year 2000 is below-average. GFEDv3 burned areas (from 1997 to October 2000) are based on relations between active fires (ATSR, VIRS, and MODIS) and mapped burned area. van der Werf et al. (2010) pointed out about 25 % of uncertainties during the years before 2001, highlighting also that the patchiness (typical of highly anthropogenically exploited areas such as the Mediterranean) may lead to an error in burned area estimates.

6.5.3 21st Century Scenarios

Finally, 21st century FEs scenarios for four species (CO, CH_4, black carbon (BC), and NO_x) in Italy across four Representative Concentration Pathways (RCPs) are illustrated in this section.

The RCPs (collaborative products between climate modelers, terrestrial ecosystem modelers and emission inventory experts) are a set of four new possible pathways of emissions and land use developments, based on consistent scenarios representative of existing literature (van Vuuren et al. 2011).

The RCPs, named according to radiative forcing target level for year 2100, included one mitigation scenario leading to a very low forcing level (RCP3PD), two medium stabilization scenarios (RCP4.5/RCP6) and one very high baseline emission scenarios (RCP8.5). Granier et al. (2011) stated that the future emissions

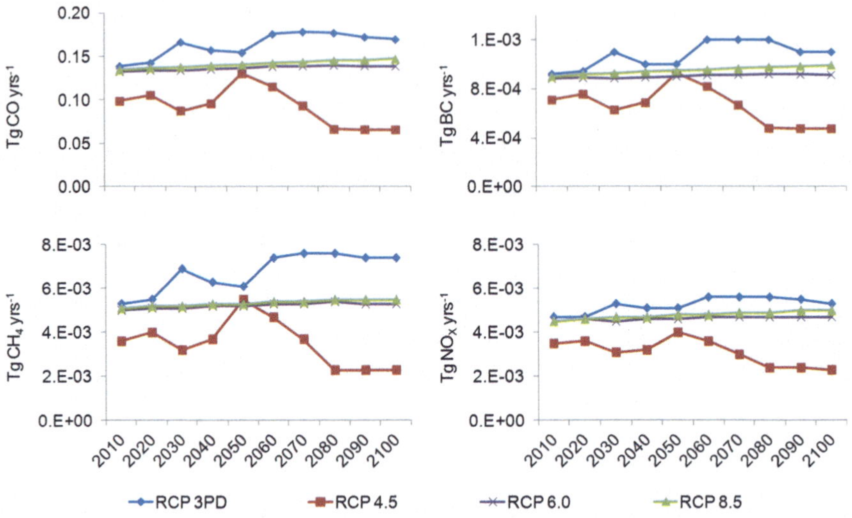

Fig. 6.4 CO (*top left*), BC (*top right*) CH_4 (*bottom left*), and NO_x (*bottom right*) emissions across the RCPs

were forced to agree with the year 2000 estimates, with the aim to guarantee continuity among emissions from past, the year 2000, and future.

Figure 6.4 showed that the trends in CO, CH_4, BC, and NO_x were quite similar. In general, RCP3PD had a peak around 2030, followed by a modest decline around 2050 and then it showed a rapid increase in 2060 followed by a stabilization. RCP6 and RCP8.5 showed a more or less stable pattern throughout the century. Finally, RCP4.5 showed the opposite trends with respect to RCP3PD, slightly decreasing around 2030 followed by a sudden change in 2050 and then as much a rapid reduction in 2080.

Emission trends are largely due to differences in the assumed climate policy, along with differences in land-use assumptions, as highlighted by van Vuuren et al. (2011). As far for land-use, which is strictly linked to forest fire pattern, RCP3PD and RCP8.5 captured an increased use of cropland, principally due to population increase. RCP4.5 and RCP6 are distinguished by a clear increasing trend in vegetation land-use (for more details, see van Vuuren et al. 2011). So that, the low emission estimates of RCP4.5, in comparison with the other scenarios, are likely due to the policies underlying this pathway, assuming that carbon in natural vegetation will be valued as part of a global climate policy (van Vuuren et al. 2011).

6.6 Conclusions

The aim of this work was to review the state of the art with reference to atmospheric emissions from forest fires, and to present historical trends and future scenarios of fire emissions, focusing on Italy. Fire emissions represent an emerging issue from local to global scale, influencing regional environment (e.g. Sinha et al. 2003; Jaffe et al. 2004; Miranda et al. 2009a, b) and global atmospheric chemistry (e.g. Novelli et al. 2003; van der Werf et al. 2004). Moreover, there is a continuous loop between fire emissions and climate, the first influencing the second while contributing to GHGs and aerosol particles, and the second affecting length and intensity of fire season. These complex interactions have been emphasized in the last years (e.g. Langmann et al. 2009), and there is an increasing interest on the definition of sound EF inventories and the regulation of regional emissions to the atmosphere (e.g. Wiedinmyer and Neff 2007).

There is a significant uncertainty regarding fire emission amount, timing, and variability (Wiedinmyer and Neff 2007). As illustrated by a number of studies (e.g. Jain 2007; Al-Saadi et al. 2008; Stroppiana et al. 2010), burned area and fuel variability are the main issues in estimating fire emissions. Langmann et al. (2009), Stroppiana et al. (2010) pointed out that, despite the improvements achieved by remote sensing, fire emissions estimates are still affected by large uncertainties, mainly due to the differences in the area burned (see Sect. 6.5). In addition, other inconsistencies derive from the assessment of emission factors and combustion estimation (Ottmar et al. 1993). Several Authors suggested that including EF large variability from field studies might improve estimate variation in tracing gas and aerosol emissions. The considerable progress in measuring methodologies and the number of projects aimed at assessing and determining accurate values of EF have been presented in Sect. 6.4, even if the amount and accuracy of EF data is not equally balanced among ecosystems (Jain 2007). Finally, the application of semi-physical modeling systems (Bacciu et al. 2012) seems to be a benefit when evaluating and assessing emissions from forest fires.

Another point of open discussion is the role of fire in climate policies. Despite the fact that long-term impacts of FE is not as much as from fossil fuel emissions, the fire impacts over shorter time periods are highly considerable (Wiedinmyer and Neff 2007). In addition, the changing climate, with a likely increase of extreme events conducive to large or more severe fires (Fried et al. 2004; Arca et al. 2009; Flannigan et al. 2009), might lead to FE that increasingly diverge from historical means. This point has been recalled also during the Durban conference in 2011: the countries, when accounting for forest management, "may exclude from the accounting emissions from natural disturbances that in any single year exceed the forest management background level" (UNFCCC 2012).

References

Albini FA (1976a) Computer-based models of wildland fire behavior: a users' manual. USDA Forest Service, Intermountain and Range Experiment Station, Ogden, UT

Albini FA, Brown JK, Reinhardt ED, Ottmar RD (1995) Calibration of a large fuel Burnout model. Int J Wildland Fire 5(3):173–192

Albini FA, Reinhardt ED (1995) Modeling ignition and burning rate of large woody natural fuels. Int J Wildland Fire 5(2):81–91

Albini FA, Reinhardt ED (1997) Improved calibration of a large fuel burnout model. Int J Wildland Fire 7(1):21–28

Al-Saadi JA, Soja AJ, Pierce RB, Szykman J, Wiedinmyer C, Emmons LK, Kondragunta S, Zhang X, Kittaka C, Schaack T, Bowman K (2008) Evaluation of near-realtime biomass burning emissions estimates constrained by satellite active fire detections. J Remote Sens Appl. doi:10.1117/1.2948785

Alves CA, Gonçalves C, Pio CA, Mirante F, Caseiro A, Tarelho L, Freitas MC, Viegas DX (2010) Smoke emissions from biomass burning in a Mediterranean shrubland. Atmos Environ 44:3024–3033

Alves CA, Vicente A, Nunes T, Gonçalves C, Fernandes AP, Mirante F, Tarelho L, Sánchez de la Campa AM, Querol X, Caseiro A, Monteiro C, Evtyugina M, Pio C (2011) Summer 2009 wildfires in Portugal: Emission of trace gases and aerosol composition. Atmos Environ. doi:10.1016/j.atmosenv.2010.10.031

Andreae MO, Merlet P (2001) Emission of trace gases and aerosols from biomass burning. Global Biogeochem Cy 15:955–966

Arca B, Bacciu V, Duce P, Pellizzaro G, Salis M, Spano D (2009) Maps of the likelihood of spread and severity of fire in a report different scenarios of fuel moisture and weather. Ital J Agrometeorology vol 14, pp 16–17. http://www.scopus.com/inward/record.url?eid=2-s2.0-79954493172&partnerID=40&md5=f3676815b79cc2565b5c46afe23d24ab

Bacciu V, Pellizzaro G, Salis M, Arca B, Duce P, Spano D (2012) Estimating vegetation fire emissions from Sardinian wildland fires (2005–2009). In: Spano D, Bacciu V, Salis M, Sirca C (eds) Modelling fire behaviour and risk. Nuova Stampa Color, pp 34–40

Barbosa PM, Grégoire JM, Pereira JMC (1999) An algorithm for extracting burned areas from time series of AVHRR GAC data applied at a continental scale. Remote Sens Environ 69:253–263

Battye W, Battye R (2002) Development of Emission Inventory Methods for Wildland Fire Final Report to US EPA. Contract # D205-01. Available at: http://www.epa.gov/ttn/chief/ap42/ch13/related/c13s01.html

Bell T, Adams M (2009) Smoke from Wildfires and Prescribed Burning in Australia: Effects on Human Health and Ecosystems. In: Bytnerowicz A, Arbaugh M, Riebau A, Andersen C (eds) Wildland fires and air pollution wildland fires and air pollution. Elsevier, Amsterdam

Bovio G (2007) Method for forest fire damage level assessment based on detectable effects. In: Ciancio O, Corona P, Marinelli M, Pettenella D (eds) Evaluation of forest fire damages in Italy. Accademia Italiana di Scienze Forestali, Florence, pp 55–60

Brown JK, Reinhardt ED, Fisher WC (1991) Predicting duff and woody fuel consumption in northern Idaho prescribed fires. For Sci 36(7):1550–1566

Burling IR, Yokelson RJ, Akagi SK, Urbanski SP, Wold CE, Griffith DWT, Johnson TJ, Reardon J, Weise DR (2011) Airborne and ground-based measurements of the trace gases and particles emitted by prescribed fires in the United States. Chem Phys Atmos. doi:10.5194/acp-11-12197-2011

Carvalho AC, Carvalho A, Martins H, Marques C, Rocha A, Borrego C, Viegas DX, Miranda AI (2011) Fire weather risk assessment under climate change using a dynamical downscaling approach. Environ Model Softw 26(9):1123–1133

Chiriacò MV, Perugini L, Cimini D, D'Amato E, Valentini R, Bovio G, Corona P, Barbati A (2013) Comparison of approaches for reporting forest fire-related biomass loss and greenhouse gas emissions in Southern Europe. Int J Wildland Fire 22(6):730–738

Clinton NE, Peng G, Klaus S (2006) Quantification of pollutants from very large wildland fires in southern California, USA. Atmos Environ 40:3686–3695

Crutzen PJ, Andreae MO (1990) Biomass burning in the tropics: Impact on atmospheric chemistry and biogeochemical cycles. Science 250:1669–1677

Crutzen PJ, Delany AC, Greenberg J, Haagenson P, Heidt L, Lueb R, Pollock W, Seiler W, Wartburg A, Zimmerman P (1985) Tropospheric chemical composition measurements in Brazil during the dry season. J Armos Chem 2:233–256

De Lauretis R, Caputo A, Dánica Cóndor R, Di Cristofaro E, Gagna A, Gonella B, Lena F, Liburdi R, Romano D, Taurino E, Vitullo M (2009) La disaggregazione a livello provinciale dell'inventario nazionale delle emissioni Anni 1990-1995-2000-2005. Rapporti 92/2009. Istituto Superiore per la Protezione e la Ricerca Ambientale

Debano LF, Neary DG, Ffolliott PF (1998) Fire's effects on ecosystems. Wiley, New York, p 333
Delmas RA et al. (1999) Experiment for regional sources and sinks of oxidants (EXPRESSO): an overview. J Geophys Res 104(D23) 30:609–624
Dennis A, Fraser M, Anderson S, Allen D (2002) Air pollutant emissions associated with forest, grassland, and agricultural burning in Texas. Atmos Environ 36:3779–3792
Dimitrakopoulos AP (2002) Mediterranean fuel models and potential fire behavior in Greece. Int J Wildland Fire. doi:10.1029/2001GL013484
EFSA (2012) Update of the monitoring of levels of dioxins and PCBs in food and feed. EFSA J 10(7):2832, 82. doi:10.2903/j.efsa.2012.2832
EC (2011) Database for Global Atmospheric Research (EDGAR) release version 42. European Commission Joint Research Centre (JRC)/PBL Netherlands Environmental Assessment Agency Emission. http://edgarjrceceuropeeu
Flannigan MD, Krawchuck MA, de Groot WJ, Wotton M, Gowman LM (2009) Implications of changing climate for global wildland fire. Int J Wildland Fire 18:483–507
Forster C et al. (2001) Transport of boreal forest fire emissions from Canada to Europe. J Geophys Res 106:22887–22906
French NHF, Goovaerts P, Kasischke ES (2004) Uncertainty in estimating carbon emissions from boreal forest fires. J Geophys Res. doi:10.1029/2003JD003635
French NHF et al (2011) Model comparisons for estimating carbon emissions from North American wildland fire. J Geophys Res. doi:10.1029/2010JG001469
Fried JS, Torn MS, Mills E (2004) The impact of climate change on wildfire severity: a regional forecast for northern California. Clim Change 64:169–191
Friedli HR, Arellano AF, Cinnirella S, Pirrone N (2009a) Mercury emissions from global biomass burning: spatial and temporal distribution. Springer, New York, pp 193–220 (Chap. 8)
Friedli H, Arellano A, Cinnirella S, Pirrone N (2009b) Initial estimates of mercury emissions to the atmosphere from global biomass burning. Environ Sci Technol 43:3507–3513
Galanter M, Levy H, Carmichael GR (2000) Impacts of biomass burning on tropospheric CO, NOx, and O_3. J Geophys Res Atmos 105:6633–6653
Granier C, Bessagnet B, Bond T, D'Angiola A, van der Gon HG, Frost G, Heil A, Kainuma M, Kaiser J, Kinne S et al (2011) Evolution of anthropogenic and biomass burning emissions at global and regional scales during the 1980–2010 period. Clim Change. doi:10.1007/s10584-011-0154-1
Hao WM, Liu MH, Crutzen PJ (1990) Estimates of annual and regional releases of CO_2 and other trace gases to the atmosphere from fires in the tropics, based on the FAO statistics for the period 1975-80. In: Fire in the tropical biota. Ecosystem processes and global challenges Goldammer JG (ed) 440–462. Ecological Studies 84, Springer, Berlin, p 497
Hao WM, Ward DE, Olbu G, Baker SP (1996) Emissions of CO_2, CO, and hydrocarbons from fires in diverse African savanna ecosystems. J Geophys Res 101:23577–23584. doi:10.1029/95jd02198
Hardy CC, Conard SG, Regelbrugge JC, Teesdale DR (1996) Smoke emissions from prescribed burning of southern California chaparral. Research Paper PNW-RP-486. Portland, OR: U.S. Department of Agriculture, Forest Service, Pacific Northwest Research Station p 37
Hardy CC, Ottmar RD, Peterson JL, Core JE, Seamon P (2001) Smoke management guide for prescribed and wildland fire: 2001 (ed) National Wildfire Coordination Group, Boise, ID, 2001. Available at http://www.treesearch.fs.fed.us/pubs/5388. Last access: 4 Aug 2011
Hodzic A, Madronich S, Bohn B, Massie S, Menut L, Wiedinmyer C (2007) Wildfire particulate matter in Europe during summer 2003: meso-scale modeling of smoke emissions, transport and radiative effects. Atmos Chem Phys 7:4043–4064
Hoelzemann JJ, Schultz MG, Brasseur GP, Granier C (2004) Global wildland fire emission model (GWEM): evaluating the use of global area burnt satellite data. J Geophys Res 109 D14S04
IPCC (2001) Climate Change 2001: The Scientific Basis. Contribution of Working Group I to the Third Assessment Report of the Intergovernmental Panel on Climate Change In: Houghton JT, Ding Y, Griggs DJ, Noguer M, van der Linden PJ, Dai X, Maskell K, Johnson CA (eds). Cambridge University Press, Cambridge, United Kingdom and New York, NY, USA, pp 881

Ito A, Penner JE (2004) Global estimates of biomass burning emissions based on satellite imagery for the year 2000. J Geophys Res Atmos 109, D14S05. doi:10.1029/2003JD004423

Hollis JJ, Matthews S, Ottmar RD, Prichard SJ, Slijepcevic A, Burrows ND, Ward B, Tolhurst KG, Anderson WR, Gould JS (2010) Testing woody fuel consumption models for application in Australian southern eucalypt forest fires. For Ecol Manag 260 (6):948–964, ISSN 0378-1127, 10.1016/j.foreco.2010.06.007

Jaffe DI, Bertschi L, Jaegle P, Novelli JS, Reid H, Tanimoto R, Vingarzan Westphal DL (2004) Long-range transport of Siberian biomass burning emissions and impact on surface ozone in western North America. Geophys Res Lett 31:L16106. doi:10.1029/2004GL020093

Jain AK (2007) Global estimation of CO emissions using three sets of satellite data for burned area. Atmos Environ 41:6931–6940. doi:10.1016/j.atmosenv.2006.10.021

Kaufman YJ, Justice C, Flynn L, Kendall J, Prins E, Ward DE, Menzel P, Setzer A (1998) Potential global fire monitoring from EOS-MODIS. J Geophys Res 103:32215–32238

Knorr W, Lehsten V, Arneth A (2011) Determinants and predictability of global wildfire emissions. Atmospheric Chemistry and Physics Discussions, submitted to The Pan European Gas-Aerosols Climate Interaction Study (PEGASOS) Special Issue

Koppmann R, Von Czapiewski K, Reid JS (2005) A review of biomass burning emissions, part I: gaseous emissions of carbon monoxide, methane, volatile organic compounds, and nitrogen containing compounds. Atmos Chem Phys Discuss 5:10455–10516

Korontzi S, Roy DP, Justice CO, Ward DE (2004) Modeling and sensitivity analysis of fire emissions in southern Africa during SAFARI 2000. Remote Sens Environ 92:255–275

Lacaux JP, Brustet JM, Delmas R, Menaut JC, Abbadie L, Bomsong B, Cachier H, Baudet J, Andreae MO, Helas G (1995) Biomass burning in the tropical savannas of Ivory 30 Coast: an overview of the field experiment fire of savannas (FOS/DECAFE 91). J Atmos Chem 22:195–216

Langmann B, Duncan B, Textor C, Trentmann J, van der Werf GR (2009) Vegetation fire emissions and their impact on air pollution and climate. Atmos Environ 43:107–116. doi:10.1016/j.atmosenv.2008.09.047

McMeeking GR, Kreidenweis SM, Baker S, Carrico CM, Chow JC, Collett JL Jr, Hao WM, Holden AS, Kirchstetter TW, Malm WC, Moosmuller H, Sullivan AP, Wold CE (2009) Emissions of trace gases and aerosols during the open combustion of biomass in the laboratory. J Geophys Res 114:D19210. doi:10.1029/2009jd011836

McRae DJ (1980) Preliminary fuel consumption guidelines for prescribed burning in Ontario slash fuel complexes. Report O-X-316. Canadian Forest Service. Great Lakes Forest Research Centre, Sault Ste. Marie, Ontario

Michel C, Liousse C, Grégoire J-M, Tansey K, Carmichael GR, Woo J-H (2005) Biomass burning emission inventory from burnt area data given by the SPOT-VEGETATION system in the frame of TRACE-P and ACE-Asia campaigns. J Geophys Res 110:D09304. doi:10.1029/2004JD005461

Mieville A, Granier C, Liousse C, Guillaume B, Mouillot F, Lamarque J-F, Grégoire J-M, Pétron G (2010) Emissions of gases and particles from biomass burning using satellite data and an historical reconstruction. Atmos Environ 44:1469–1477. doi:10.1016/j.atmosenv.2010.01.011

Miranda AI, Marchi E, Ferretti M, Millan MM (2009a) Forest fires and air quality issues in Southern Europe. In: Bytnerowicz A, Arbaugh M, Riebau A, Andersen C (eds) Wildland fires and air pollutionwildland fires and air pollution. Elsevier, Amsterdam

Miranda AI, Borrego C, Martins H, Martins V, Amorim JH, Valente J, Carvalho A (2009b) Forest fire emissions and air pollution in southern Europe. In: Chuvieco E (ed) Earth observation of wildland fires in Mediterranean ecosystems. Springer, Berlin

Miranda AI, Coutinho M, Borrego C (1994) Forest fire emissions in Portugal: A contribution to global warming? Environ Pollut 83(1–2):121–123

Miranda AI, Ferreira J, Valente J, Santos P, Amorim JH, Borrego C (2005a) Smoke measurements during Gestosa-2002 experimental field fires. Int. J. Wildland Fire 14:107–116

Miranda AI, Borrego C, Sousa M, Valente J, Barbosa P, Carvalho A (2005b) Model of forest fire emissions to the atmosphere. Deliverable D252 of SPREAD Project (EVG1-CT-2001-00043). AMB-QA-07/2005. University of Aveiro, Aveiro, Portugal

Miranda AI, Martins H, Carvalho A, Borrego C (2005c) Modelling smoke effects on Lisbon air quality. In: Proceedings in CD-Rom Sixth fire and forest meteorology symposium and the 19th Interior West Fire Council meeting, Oct 25–27. Canmore, AB, Canada

Mouillot F, Field CB (2005) Fire history and the global carbon budget: a 1°x1° fire history reconstruction for the 20th century. Global Change Biol 11(3):398–420

Naeher LP, Brauer M, Lipsett M, Zelikoff JT, Simpson CD, Koenig JQ, Smith KR (2007) Woodsmoke health effects: a review. Inhal Toxicol 19(1):67–106

Narayan C, Fernandes PM, van Brusselen J, Schuck A (2007) Potential for CO_2 emission mitigation in Europe through prescribed burning in the context of the Kyoto Protocol. Forest Ecol Manag 251:164–173

Novelli PC, Masarie KA, Lang PM, Hall BD, Myers RC, Elkins JW (2003) Reanalysis of tropospheric CO trends: Effects of the 1997–1998 wildfires. J Geophys Res Atmos 108(D15):4464. doi:10.1029/2002JD003031

Ottmar RD, Miranda AI, Sandberg DV (2009) Characterizing sources of emissions from wildland fires. In: Bytnerowicz A, Arbaugh M, Riebau A, Andersen C (eds) Wildland fires and air pollutionwildland fires and air pollution. Elsevier, Amsterdam

Ottmar RD, Prichard SJ, Vihnanek RE, Sandberg DV, Bluhn A (2006) Modification and validation of fuel consumption models for shrub and forested lands in Southwest, Pacific northwest, Rockies, Midwest, Southeast and Alaska. Final Report JFSP Project 98-1-9-06. USDA Forest Service, fire and environmental applications team, Pacific Northwest Research Station, Pacific Wildland Fire Sciences Laboratory, Seattle, WA

Ottmar RD, Burns MF, Hall JN, Hanson AD (1993) Consume Version 1.0. General Technical Report PNW GTR 304. USDA Forest Service, Pacific Northwest Research Station

Peterson JL (1987) Analysis and reduction of the errors of predicting prescribed burn emissions. Master's thesis, University of Washington, Seattle, p 70

Peterson J, Sandberg D (1988) A national PM10 inventory approach for wildfires and prescribed fires. In: Mathai C, Stonefield D (eds) Proceedings of the PM10 implementation of standards: an APCA/EPA international specialty conference, San Francisco, CA

Prichard SJ, Ottmar RD, Anderson GK (2005) Consume 3.0 User's Guide. United States Department of Agriculture Forest Service, Pacific Northwest Research Station, Seattle, WA

Radke LF, Hegg DA, Hobbs PV, Nance JD, Lyons JH, Laursen KK, Weiss RE, Riggan PJ, Ward DE (1991) Particulate and trace gas emissions from large biomass fires in North America. In: Levine JS (eds) Global biomass burning: atmospheric, climatic, and biospheric implications, MIT Press, Cambridge, MA, pp 209–224

Randerson JT, Liu H, Flanner MG, Chambers SD, Jin Y, Hess PG, Pfister G, Mack MC, Treseder KK, Welp LR, Chapin FS, Harden JW, Goulden ML, Lyons E, Neff JC, Schuur EAG, Zender CS (2006) The impact of boreal forest fire on climate warming. Science 314:1130–1132

Reid JS, Eck TF, Christopher SA, Koppmann R, Dubovik O, Eleuterio DP, Holben BN, Reid EA, Zhang J (2005) A review of biomass burning emissions part III: intensive optical properties of biomass burning particles. Atmos Chem Phys 5:827–849

Reinhardt ED, Keane RE, Brown JK (1997) First order fire effects model: FOFEM 4.0, User's Guide. General Technical Report INT-GTR-344. USDA Forest Service, Intermountain Research Station, Ogden, UT

Schollnberger H, Aden H, Scott BR (2002) Respiratory tract deposition efficiencies: evaluation of effects from smoke released in the Cerro Grande forest fire. J Aerosol Med 15:387–399

Schultz MG, Heil A, Hoelzemann JJ, Spessa A, Thonicke K, Goldammer JG, Held AC, Pereira JMC, van het Bolscher M (2008) Global wildland fire emissions from 1960 to 2000. Global Biogeochem Cycles 22, GB2002. doi:10.1029/2007GB003031

Seiler W, Crutzen P (1980) Estimates of gross and net fluxes of carbon between the biosphere and the atmosphere from biomass burning. Climatic Change 2:207–247

Shea RW, Shea BW, Kauffman JB, Ward DE, Haskins CI, Scholes MC (1996) Fuel biomass and combustion factors associated with fires in savanna ecosystems of South Africa and Zambia. J Geophys Res 101:23551–23568

Simmonds PG, Manning AJ, Derwent RG, Ciais P, Ramonet M, Kazan V, Ryall D (2005) A burning question. Can recent growth rate anomalies in the greenhouse gases be attributed to large-scale biomass burning events? Atmos Environ 39:2513–2517

Sinha P, Hobbs PV, Yokelson RJ, Bertschi IT, Blake DR, Simpson IJ, Gao S, Kirchstetter TW, Novakov T (2003) Emissions of trace gases and particles from savanna fires in southern Africa. J Geophys Res 108:8487. doi:10.1029/2002jd002325

Spichtinger N, Wenig M, James P, Wagner T, Platt U, Stohl A (2001) Satellite detection of a continental-scale plume of nitrogen oxides from boreal forest fires. Geophys Res Lett 28:4579–4582

Stroppiana D, Brivio PA, Grégoire J-M, Liousse C, Guillaume B, Granier C, Mieville A, Chin M, Pétron G (2010) Comparison of global inventories of CO emissions from biomass burning derived from remotely sensed data. Atmos Chem Phys 10:12173–12189. doi:10.5194/acp-10-12173-2010

Swap RJ, Annegarn HJ, Suttles JT et al. (2003) Africa burning: a thematic analysis of the Southern African Regional Science Initiative (SAFARI 2000). J Geophys Res 108(D13), 30:8465. doi:10.1029/2003JD003747

Thonicke K, Spessa A, Prentice IC, Harrison SP, Dong L, Carmona-Moreno C (2010) The influence of vegetation, fire spread and fire behaviour on biomass burning and trace gas emissions: results from a process-based model. Biogeosciences 7:1991–2011. doi:10.5194/bg-7-1991-2010

UNFCCC (2012) Report of the conference of the parties serving as the meeting of the parties to the Kyoto Protocol on its seventh session, held in Durban from 28 November to 11 December 2011. Part Two: Action taken by the Conference of the Parties serving as the meeting of the Parties to the Kyoto Protocol at its seventh session FCCC/KP/CMP/2011/10/Add.1

Urbanski SP, Hao WM, Baker S (2009) Chemical composition of wildland fire emissions. In: Bytnerowicz A, Arbaugh M, Riebau A, Andersen C (eds) Wildland fires and air PollutionWildland fires and air pollution. Elsevier, Amsterdam

Urbanski SP, Hao WM, Nordgren B (2011) The wildland fire emission inventory: western United States emission estimates and an evaluation of uncertainty. Atmos Chem Phys 11:12973–13000. doi:10.5194/acp-11-12973-2011

van der Werf GR, Randerson JT, Collatz GJ, Giglio L, Kasibhatla PS, Arellano AF, Olsen SC, Kasischke ES (2004) Continental-scale partitioning of fire emissions during the 1997 to 2001 El Nino/La Nina period. Science 303:73–76

van der Werf GR, Randerson JT, Giglio L, Collatz GJ, Kasibhatla PS, Arellano AF (2006) Interannual variability in global biomass burning emissions from 1997 to 2004. Atmos Chem Phys 6:3423–3441. doi:10.5194/acp-6-3523-2006

van der Werf GR, Randerson JT, Giglio L, Collatz GJ, Mu M, Kasibhatla PS, Morton DC, DeFries RS, Jin Y, van Leeuwen TT (2010) Global fire emissions and the contribution of deforestation savanna forest agricultural and peat fires (1997–2009). Atmos Chem Phys 10:11707–11735. doi:10.5194/acp-10-11707-2010

van Vuuren DP, Edmonds JA, Kainuma M, Riahi K, Thomson AM, Hibbard K, Hurtt GC, Kram T, Krey V, Lamarque J-F, Masui T, Meinshausen M, Nakicenovic N, Smith SJ, Rose S (2011) The representative concentration pathways: an overview. Clim Change 109:5–31. doi:10.1007/s10584-011-0148-z

Ward DE, Hardy CC (1991) Smoke emissions from wildland fires. Environ Int 17:117e–134

Ward DE, Hao WM, Susott R, Babbitt R, Shea R (1996) Effect of fuel composition on combustion efficiency and emission factors for African savanna ecosystems. J Geophys Res 101(D19):23569–23576. doi:10.1029/95JD02595

Wiedinmyer C, Neff JC (2007) Estimates of CO_2 from fires in the United States: implications for carbon management. Carbon Balance Manage 2:10. doi:10.1186/1750-0680-2-10

Wiedinmyer C, Akagi SK, Yokelson RJ, Emmons LK, Al-Saadi JA, Orlando JJ, Soja AJ (2010) The fire inventory from NCAR (FINN)—a high resolution global model to estimate the emissions from open burning. Geosci Model Dev Discuss 3:2439–2476. doi:10.5194/gmdd-3-2439-2010

Yokelson RJ, Karl T, Artaxo P, Blake DR, Christian TJ, Griffith DWT, Guenther A, Hao WM (2007) The tropical forest and fire emissions experiment: overview and airborne fire emission factor measurements. Atmos Chem Phys 7:5175–5196. doi:10.5194/acp-7-5175-2007

Chapter 7
Carbon Losses Due to Wood Harvesting and the Role of Wood Products

Marco Marchetti, Gherardo Chirici and Bruno Lasserre

Abstract The carbon stock in wood and paper products is increasing in Italy, and the same trend is expected in the coming decades. In forest ecosystems firewood and forest harvesting represent a net carbon loss but the use of wood, a carbon-neutral renewable resource for generating energy also has a strong substitution effect as it avoids the use of fossil fuels which are highly CO_2 emitting. The use of wood for construction purposes, substituting traditional materials, tends to increase carbon sequestration and contributes to climate change mitigation. The application of the GHGs accounting methods (IPCC 2003) suffers in Italy both for the lack of accuracy of wood harvesting official statistics and for the high level of uncertainty in the definition of wood products lifespan. Many authors have demonstrated a large underestimation of clear cuts areas in Italy leading to an underestimation of carbon loss due to a harvesting of about 2 Mt annually. However, it has been recently proved that multitemporal high resolution remotely sensed images may be operatively used with a probabilistic sampling procedure to obtain a more reliable estimation of annual wood harvesting extents. In any case an increase of the use of wood products for energy, building and furniture purposes may contribute to the reduction of GHG emissions and to a more sustainable development.

7.1 Introduction

In the last decades the need for the development of valuable measures for the mitigation of GHGs emissions and their effect on climate change significantly influenced research in the forest sector. In fact, forests may significantly contribute in

M. Marchetti · G. Chirici · B. Lasserre (✉)
Department of Biosciences and Territory, University of Molise, Pesche, IS, Italy
e-mail: lasserre@unimol.it

R. Valentini and F. Miglietta (eds.), *The Greenhouse Gas Balance of Italy*,
Environmental Science and Engineering, DOI 10.1007/978-3-642-32424-6_7

stabilizing and potentially decreasing the concentration of carbon dioxide in the atmosphere. However, natural or human-induced disturbances, such as wood harvesting, forest fires or land use changes, may dramatically alter these processes determining a consistent release of GHGs back in the atmosphere.

For instance, the global Forest Resource Assessment (FRA) 2010 reported that global wood removals in 2005 amounted to 3.4×10^9 m^3 (accounting for 0.7 % of growing stock), roughly corresponding to 8 Gt CO_2 equivalent, of which about half were industrial roundwood and half woodfuel. Furthermore, considering that informally and illegally removed wood, especially woodfuel, is not usually recorded, the actual amount of wood removals is undoubtedly higher (FAO 2010a).

Monitoring and assessing wood harvesting is relevant not only for quantifying forest carbon budget but also to evaluate the sustainability of forest management. For instance, countries are asked to report this information within the framework of Forest Europe (the former Ministerial Convention on the Protection of Forests in Europe) (MCPFE 2009), in the Montréal Process (2005), in the framework of the UNFCCC (2007) and the Kyoto Protocol (1997) and for the global FRA carried out by the Food and Agriculture Organization of the United Nations (FAO 2010a).

According to Mader (2007), the forest sector's carbon cycle is structured in four major carbon pools—the forest, forest products-in-use, products disposed in landfills and fossil fuel displaced by forest products and bioenergy in end-use markets.

Carbon stored in forest products ensures a substantial degree of permanence in the carbon storage and dampens flow back to the atmosphere. Moreover, the global carbon balance largely benefits from using wood products as substitutes for other materials requiring much more energy to be produced. In addition, when forest biomass is used to produce energy, replacing fossil fuel, it permanently offsets the carbon emissions from displaced fossil fuels. When product substitution is considered, wood products can contribute to a significant reduction in the atmospheric greenhouse emissions by generating bioenergy and displacing fossil fuel-intensive products.

According to FAO (2010b), when looking at the emission, sequestration and substitution accomplished in the value chain, the use of wood products could avoid global emissions in the atmosphere of over 150 Mt CO_2 equivalent per year. Nabuurs et al. (2007) estimate that the energy derived from forest biomass could reduce global emissions by 400 Mt–4.4 Gt CO_2 equivalent per year.

Currently, the European Union is again in the driver's seat in the implementation of a realistic policy of emission reductions and towards the closing of the important gap of unaccounted CO_2 savings and emissions from forest-related activity. Up until the Durban climate conference (UNFCCC), reporting assumed the instant oxidation of all harvested biomass. HWP is now mandatory and can be accounted based either on the current instant oxidation approach, or based on the Production approach. However, HWP leading to deforestation will be counted as instant oxidation. Accounting for HWP has the potential to enhance the accuracy of reporting when and where emissions occur and to further enhance incentives to substitute HWP products for more carbon intensive complements (e.g. steel, cement), thus reducing GHG-emissions.

7.2 Forest Harvesting in Italy

7.2.1 Official Statistics

In Italy forests cover about 87,592 km^2 (29 % of the total land area), of which 42 % is coppice forest (INFC 2007). The main sylvicultural system adopted is coppice with standards, applied with a rotation period varying between 15 and 35 years. Stands are thus even-aged, made by 40–200 trees per hectare with an age varying from two to three times the rotation period (Ciancio et al. 2006). At the end of the rotation period, harvesting activities are performed by clearcutting between November and April. This type of sylviculture mainly provides wood for energy even if it does not succeed in satisfying the internal consumption.

Small clearcuts in coppice forests (for areas generally under 3–10 ha, depending of local laws), simply has to be communicated to the Regional authority and to the Italian Forest Service (Corpo Forestale dello Stato—CFS) by the forest owner. For larger clearcuts a formal harvesting project has to be approved by the Regional authority. Both communications and authorizations data are collected, analyzed and published by ISTAT.

In Fig. 7.1, data from National Statistics Agency (Istituto Nazionale di Statistica—http://agri.istat.it—visited 07/2012) on the total volume of wood harvested from forests are drawn for the years from 1999 to 2010. Data is reported for the whole Italian area and for five macro-regions. Throughout these years, the average harvested volume from forest at national level amounts to almost 7,700,000 m^3 y^{-1} (accounting for 0.5 % of growing stock), of which more or less 68 % is fuelwood.

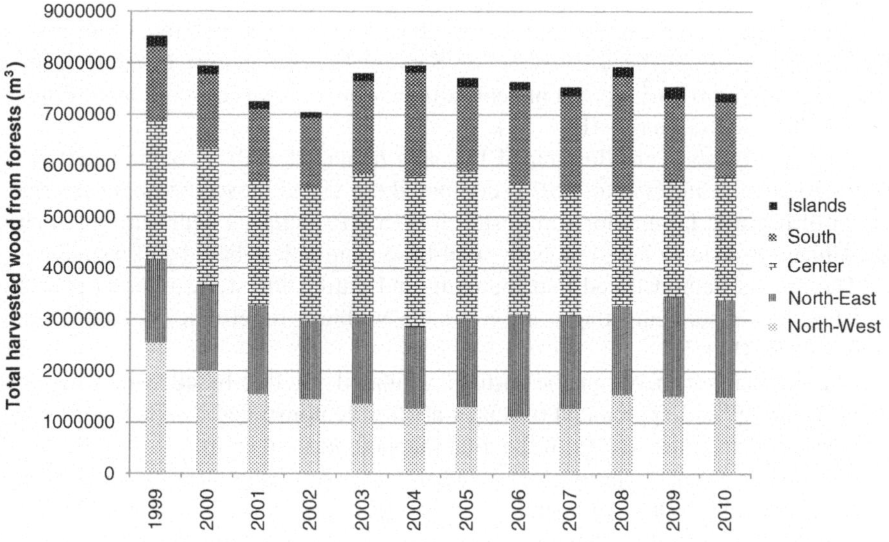

Fig. 7.1 Temporal trends of wood harvested from forests in Italy (*data source* ISTAT)

However, since resulting forest wood harvesting aggregated statistics apparently underestimate real wood production, the whole reporting system was frequently criticized in literature (Cutolo 2000; Corona et al. 2004). Forest owners may in fact communicate clearcuts smaller than the effective ones avoiding the cost of a formal harvesting project and simply cutting in a larger area. Just a few clearcuts are verified in the field by local authorities.

It is evident that since updated forest wood harvesting statistics are so important in supporting strategic forest planning actions, an accurate, cheap and feasible monitoring system should be implemented. The integrated use of remote sensing technologies should be considered. The results recently published in Chirici et al. (2011) are here synthesized.

7.2.2 Remote Sensing for Harvesting Monitoring

Monitoring of forest disturbances by remote sensing is usually implemented through the use of change detection techniques (e.g. Coppin et al. 2004; Lu et al. 2004).

Forest harvesting monitoring through satellite image analysis is reported for the tropics (e.g. Souza et al. 2005), USA (e.g. Schroeder et al. 2007), Canada (e.g. Wulder et al. 2007), and boreal Europe (e.g. Heikkonen and Varjo 2004).

Chirici et al. (2011) presented the first study specifically oriented to satellite monitoring of wood harvesting in coppice forests. The authors tested manual and semi-automatic methods to define a correction coefficient to be applied to official statistics in order to obtain a more reliable data on annual harvested extents.

The test was carried out within the framework of the project Global Monitoring for Environment and Security (GMES) Service Element Forest Monitoring in a test area of 34,000 km^2 located in central Italy. The area was covered by multitemporal SPOT5 HRG images (Fig. 7.2).

Coppice forests were dominated by oaks managed with a rotation period of about 20 years. Chirici et al. (2011) compared the visual on-screen photo-interpretation of infrared false color composite SPOT5 HRG images with semi-automatic classification systems based on both pixel-based and object-based approaches.

The results demonstrated that mapping and dating forest clearcuts is possible and feasible if post-cut images are available within a timeframe of 1 or 2 years since the cut (Fig. 7.3).

The comparison between the in-field validated satellite-based map with over 9,500 clearcuts and aggregated official statistics demonstrated a significant underestimation by the latter (65 % of the total mapped clearcut area).

The object-oriented approach resulted in better results confirming the findings of Desclée et al. (2006). An operative probabilistic procedure based on VHR (very high resolution) satellite samples was proposed by the authors to correct official harvesting statistics.

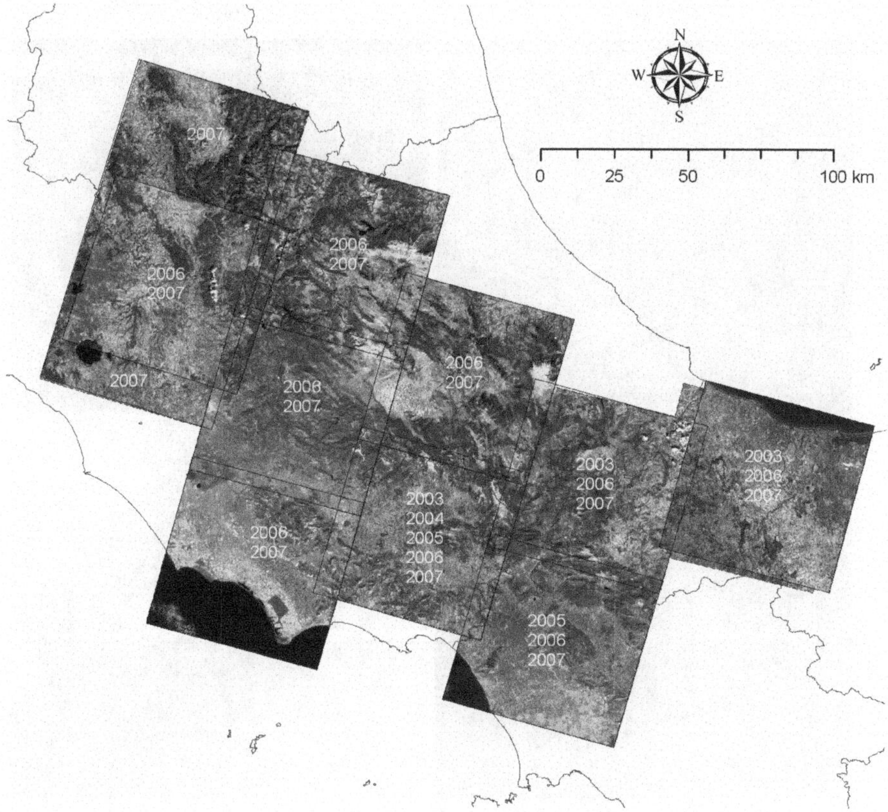

Fig. 7.2 A mosaic of some of the SPOT5 HRG images (band 4 in the near infrared channel). For each area the dates of the images used in the study are also reported (from Chirici et al. 2011 modified)

7.2.3 Carbon Loss Due to Harvesting

According to Papale et al. (2005), the losses of carbon ($C_{\text{wood utilization}}$) from forests due to forest harvesting can be assessed as follows:

$$C_{wood\ utilization} = HV \cdot k_1 \cdot bd \cdot k_2[\text{t}] \tag{7.1}$$

where

HV = harvested wood volume [m^3];

k_1 = average expansion coefficient to take into account the wood residuals left in the forest stands (cutting woody debris); this coefficient was set equal to 1.176, according to Corona and Nocentini (2002);

bd = average wood basal density of forest tree species in Italy, set equal to 0.65 t m^{-3}, according to APAT (2003);

k_2 = average carbon content per unit of wood biomass of forest tree species [t t^{-1}], set equal to 0.5 according to APAT (2003).

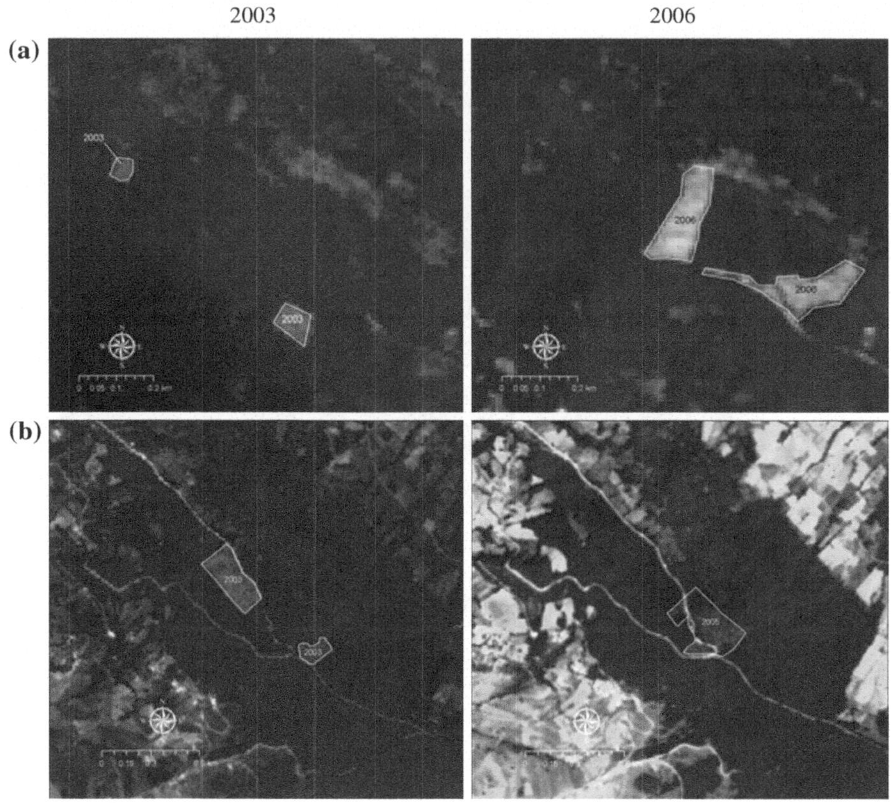

Fig. 7.3 Examples of comparison of SPOT5 HRG infrared images from different years evidencing clearcut areas (*white delineated polygons* with the clearcut year). The images were acquired in late spring or summer of 2003 and 2006. Both cases **a** and **b** are from Regione Molise (from Chirici et al. 2011 modified)

According to ISTAT, 54 % of annual harvested volume is represented by coppice forests, corresponding to 4.26×10^6 m^3. Thus wood removals through clear cuts in Italy officially represent an annual loss of carbon of about 6 Mt CO_2.

As a conclusion, on the basis of the underestimation of official statistics on clear cuts harvesting presented above, it can be hypothesized an underestimation on carbon losses at national level of about 2 Mt CO_2.

7.3 Wood Products

After forest harvesting, wood products continue to store carbon for a certain time, from some weeks for paper to many years for timber beam used in construction. At the end of their life cycle, wood products can be recycled, stored in landfills or used for generating energy, thus substituting fossil fuels. At the same time various types of carbon storage and new raw material are provided by forests for industrial purposes. After a highly variable period of time, wood products will release carbon back to the atmosphere through emissions from production processes and decay of wood-based products (Eggers 2002). Research on this topic is still novel and a substantial effort was given in the framework of Cost Action E31 "Management of Recovered Wood" to reduce the uncertainties in determining the lifespan of wood material after harvesting.

Carbon is stocked in wood and paper products in use and in landfills. The global pool of carbon stored in forests products is estimated to be growing (Fig. 7.4), from 2007, by 150 Mt C (540 Mt CO_2 equivalent) per year (Miner and Perez-Garcia 2007) reaching 5 Gt C (18.3 Gt CO_2 equivalent) in 2010. For this reason wood and paper products may potentially play an important role in mitigating carbon emissions in the atmosphere. This topic was investigated in several studies. Skog et al. (2000) used historical data and long-range projections in order to track roundwood and carbon disposition through to end uses such as housing and paper. Miner and Perez-Garcia (2007) examine the significance of emissions, sequestration and avoided emissions of various wood products in order to assess the carbon profile of the forest product industry.

Some studies assess carbon sequestration for a range of hypothetical conditions of forest growth, harvest, end use, and discarding (Schlamadinger and Marland 1996).

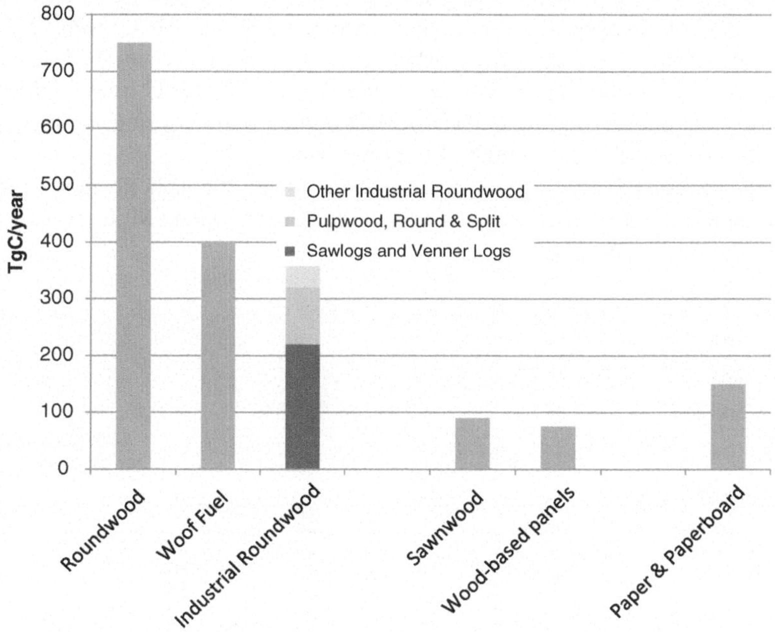

Fig. 7.4 Estimated C fluxes in global production of harvested wood products in 2003 (based on FAO STAT 2005)

A worldwide study by Winjum et al. (1998) estimates net flows of carbon out of forests and into products by using two accounting frameworks—the stock change method, and the atmospheric flow method. They use simplified assumptions to make estimates of net stock changes and net emissions to the atmosphere by world regions.

The highly variable duration of carbon stock in wood products depends on their composition, their level of production-consumption and their average life cycle, an extremely variable parameter (from few months for newsprint to many decades for laminated beam).

7.3.1 Carbon Accounting Scheme for Wood Products in Italy

The question on how to account emissions or stock-changes for HWP in the context of the UNFCCC has been extensively discussed and assessed internationally. Different approaches have been proposed and they differ in how they allocate emissions between wood producing and consuming countries, and in what processes they focus on (Brown et al. 1998; Winjum et al. 1998; Lim et al. 1999).

In a study performed by Kloehn and Ciccarese (2005) the three approaches proposed by IPCC (2003) were applied in order to estimate GHG balances in wood products in Italy: production approach, stock change approach and atmospheric flow approach (Nabuurs et al. 2003; Pingoud et al. 2006).

According to Kloehn and Ciccarese (2005), in 2008 in Italy, the stock was estimated in 59 Mt C, using the production approach and 140 Mt C using the stock change approach (Fig. 7.5). The carbon sink was estimated in 0.25 Mt C (or 0.92 Mt of CO_2) using the production approach and 1.12 Mt C (or 4.11 Mt CO_2) using the stock change approach. The lowest value obtained using the production approach reflects the large volume of imported wood.

The results of this study indicate that through the atmospheric flow approach wood products in Italy become a net source of carbon, up to 0.46 Mt C (or 1.69 Mt of

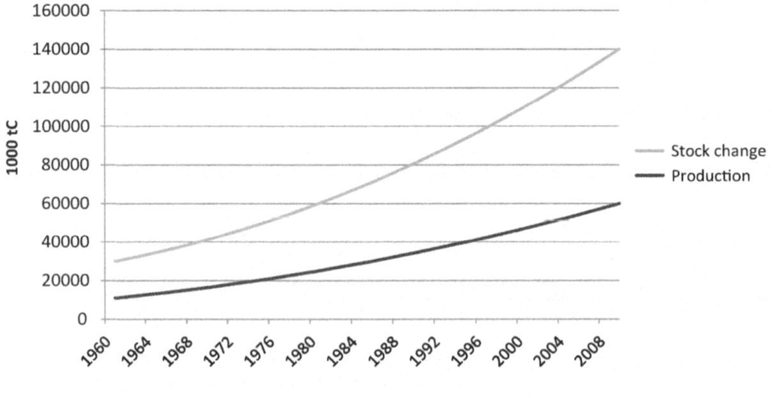

Fig. 7.5 Trend of carbon stock in wood products (from Kloehn and Ciccarese 2005 modified)

CO_2). This method tends to limit the possibility of properly accounting carbon sinks in those countries where, like Italy, timber is imported to produce furniture and other finished wood products which are then exported abroad.

Moreover, a contrast of the stock change approach versus the production approach resulted in very different figures in the landfills carbon stock estimation (49 % of the total against 63 % respectively).

At the moment carbon stock estimations in Italy are based on assumptions and data that should be better tested, also in light of the poor quality of the above mentioned statistics on wood harvesting (Corona et al. 2004) and imports of timber in Italy (Pettenella and Ciccarese 2009).

On the basis of the experience carried out by Kloehn and Ciccarese (2005), the inclusion of forest products in the national GHGs accounting system could dramatically affect the final Italian balance depending on the adopted accounting method.

However, at the moment, as clearly demonstrated by Hashimoto (2008), none of the accounting approaches can provide incentives which are consistent and compatible with the overall objectives of forest and environmental policies such as the promotion of forest conservation, promotion of energy use of wood, promotion of material use of wood, promotion of recovery of wood. The best approach in order to achieve each policy goal actually varies, and the best approaches for specific policy goals might present problems for others.

7.3.2 Improvement of Wood Products Uses

Wood is an excellent construction material because of its physical characteristics. It is a low density material with a high carrying capacity, as well as a high thermic and acoustic efficiency, very suitable for constructions in seismic areas for its high energy dissipation capacity. Moreover, wood is a renewable material requiring limited external energy to be produced and transformed. It can therefore be used in place of many other materials whose production require a much higher level of energy, such as iron and aluminum, thereby reducing emissions of GHGs (FAO 2010b). It is therefore relevant to identify practical solutions and valuable techniques to increase the use of wood in domestic and industrial applications replacing other materials.

Some studies related to the energy required to build up constructions with different combinations of materials suggest that maximizing the use of wood in new buildings could reduce in a significant way GHGs emissions generated in the production of building materials (Skog and Nicholson 1998). For example, the energy needed for the production of an iron beam exceeds up to nine times the energy needed to produce a wood product with the same length and carrying capacity (Burschel and Kürsten 1992). Gustavsson et al. (2006), comparing two multi-storey buildings, one made of wood and the other of concrete, show that the CO_2 emissions for the construction with wood were significantly lower than those needed for the construction with concrete, mainly due to the much higher emissions for the production of this latter.

A study in Sweden shows that the reduced gas emissions to avoid the use of traditional construction materials in favor of wood, compensate the temporary carbon losses stocked in forest due to wood harvesting (Eriksson et al. 2007). A similar conclusion is presented by Taverna et al. (2007) for Switzerland: a more intensive timber harvesting in national forests for construction uses would determine a decrease in GHGs emissions of up to 13 %.

Furthermore, wood residues from forest harvesting or those obtained from construction wood recycling may be used for producing energy in substitution of fossil fuels and thus significantly contributing to the reduction of CO_2 emissions.

According to the United Nation's Environment Programme's Sustainable Building and Climate Initiative, while all stages of a building's life-cycle (including construction and demolition) produce GHG emissions, a building's operational phase accounts for 80–90 % of the emissions resulting from energy use. The use of woody biomass for domestic heating can therefore play an important role in the emission budgets.

Several tools could be used so to improve the use of wood product.

A carbon tax applied to traditional construction materials proportional to the amount of gas emitted for their production could stimulate the use of wood products.

The certification of forest products may also increase the demand for wood, informing consumers about sustainable forest management applied for wood production.

Wood materials should be promoted informing consumers about their technical and environmental benefits.

Even the publicity of prestigious wood constructions can be used to successfully disseminate these information. In this sense, a positive case is the large roof built in Hanover for the Expo 2000 (http://www.krusi.com/Expo2000.html), or the large echo given to the construction of a 16-storey wooden building in Norway (http://inhabitat.com/worlds-tallest-wooden-building-planned-for-norway/).

7.4 Conclusions

The results from Chirici et al. (2011) confirmed a 35 % underestimation of official wood harvesting statistics from coppice forests in Italy originally hypothesized by Cutolo (2000) and Corona et al. (2004). Multitemporal high resolution remotely sensed images can be operatively used with a probabilistic sampling procedure to obtain a more reliable data on annual wood harvesting extents. This underestimation of clear cuts area brings to an underestimation of carbon loss due to harvesting of about 2 Mt annually.

The use of wood for construction purposes, substituting traditional materials, tends to increase carbon sequestration and to contribute to climate change mitigation. The use of wood for generating energy has also a strong substitution effect as it avoids using fossil fuels known as highly CO_2 emitting.

In addition, forest harvesting for wood removals surely contributes to the limitation of Carbon losses due to biotic and abiotic disturbances such as forest fires, pest and diseases.

In Italy, the carbon stock in wood products is growing and the same trend is expected in the coming decades even if, in a longer perspective, an equilibrium will be probably reached (Ciccarese and Kloehn 2010).

In Italy, the application of the GHGs accounting methods (IPCC 2003) suffers both for the lack of accuracy of wood harvesting official statistics and for the high level of uncertainty in the definition of wood products lifespan.

Accounting, reporting, and policy challenges exist both nationally and internationally to determine how to report, credit and structure policy to provide forest management incentives that result in appropriate use of wood product C-storage or the substitution effect as GHG mitigation/management strategies. The challenges are complex (UNFCC 2003), but must be dealt with in order to allow market forces to properly recognize the value of forests in the reduction of GHG emissions.

In any case, regardless of the decisions of the UNFCCC on harvested wood products accounting, increasing the use of wood products for energy and building purposes may contribute to the reduction of GHG emissions and to a more sustainable development. It is therefore essential to increase dissemination actions informing potential customers of technical and environmental advantages of wood materials.

References

APAT (2003) Le biomasse legnose. A cura di Ciccarese L, Spezzati E, Pettenella D. Rapporti 30, Roma

Brown S, Lim B, Schlamadinger B (1998) Evaluating approaches for estimating net emissions of carbon dioxide from forest harvesting and wood products. Intergovernmental Panel on Climate Change, Meeting Report, Dakar, Senegal. IPCC/OECD/IEA, Paris, 5–7 May 1998

Burschel and Kürsten (1992) Wald und Forstwirtschaft im Kohlenstoffhaushalt der Erde. In: Produktionsfaktor Umwelt: Klima - Luft. Hrsg. Verbindungsstelle Landwirtschaft-Industrie e.V., Energiewirtschaft und Technik Verlagsges. mbH, Düsseldorf, 97–125

Chirici G et al (2011) Large-scale monitoring of coppice forest clearcuts by multitemporal very high resolution satellite imagery. A case study from central Italy. Remote Sens Environ 115(4):1025–1033

Ciancio O, Corona P, Lamonaca A, Portoghesi L, Travaglini D (2006) Conversion of clearcut beech coppices into high forests with continuous cover: a case study in central Italy. For Ecol Manage 3:235–240

Ciccarese L, Kloehn S (2010) La capacità fissativa di carbonio nei prodotti legnosi: una stima per l'Italia. Agriregionieuropa, 6(21)

Coppin P, Jonckheere I, Nackaerts K, Muys B (2004) Digital change detection methods in ecosystem monitoring: a review. Int J Remote Sens 10:1565–1596

Corona P, Nocentini S (2002) Biomasse e residui legnosi dei boschi italiani. Biomasse agricole e forestali a uso energetico, Agra Editrice, Roma 24–34

Corona P, Macrì A, Marchetti M (2004) Boschi e foreste in Italia secondo le più recenti fonti informative. L'Italia Forestale e Montana 2:119–136

Cutolo N (2000) Il progetto per il settore forestale. I riferimenti delle nuove statistiche forestali: integrazione nel sistema e ottica di filiera. Istat, Servizio Agricoltura, Roma, pp 16–27

Desclée B, Bogaert P, Defourny P (2006) Forest change detection by statistical object-based method. Remote Sens Environ 102:1–11

Eggers T (2002) The impacts of manufacturing and utilisation of wood products on the European carbon budget. EFI Internal Report No. 9, 90 p

Eriksson E, Gillespie A, Gustavsson L, Langvall O, Olsson M, Sathre R, Stendahl J (2007) Integrated carbon analysis of forest management practices and wood substitution. Can J For Res 37:671–681

FAO (2010a) Global forest resources assessment, 2010-Main report. FAO Forestry Paper 163, Rome

FAO (2010b) Impact of the global forest industry on atmospheric greenhouses gases, Reid Miner, FAO Forestry Paper 159, Rome

Gustavsson L, Pingoud K, Sathre R (2006) Carbon dioxide balance of wood substitution: comparing concrete and wood framed buildings. Mitig Adapt Strat Glob Change 11(3):667–691

Hashimoto S (2008) Different accounting approaches to harvested wood products: their incentives to achievement of major policy goals. Environ Sci Policy 11:756–771

Heikkonen J, Varjo J (2004) Forest change detection applying landsat thematic mapper difference features: a comparison of different classifiers in boreal forest conditions. Forest Sci 50:579–588

INFC (2007) Le stime di superficie 2005 – Prima parte. In: Autori G, Tabacchi F, De Natale, L Di Cosmo, Floris A, Gagliano C, Gasparini P, Genchi L, Scrinzi G, Tosi V. Inventario Nazionale delle Foreste e dei Serbatoi Forestali di Carbonio. MiPAF – Corpo Forestale dello Stato - Ispettorato Generale, CRA - ISAFA, Trento. http://www.infc.it

IPCC (2003) Penman J, Gytarsky M, Hiraishi T, Krug T, Kruger D, Pipatti R, Buendia L, Miwa K, Ngara T, Tanabe K, Wagner F, Good Practice Guidance for Land Use, land-Use Change and Forestry IPCC/IGES, Hayama, Japan

Kloehn S, Ciccarese L, (2005). Applying the IPCC GPG for LULUCF approaches for assessing changes in carbon stocks and emissions of green-house gas for harvested wood products in Italy. Available at www.apat.gov.it

Lim B, Brown S, Schlamadinger B (1999) Carbon accounting for forest harvesting and wood products: review and evaluation of different approaches. Environ Sci Policy 2(2):207–216

Lu D, Mausel P, Brondizio E, Moran E (2004) Change detection techniques. Int J Remote Sens 25:2365–2401

Mader S (2007) Climate project: review and compare carbon accounting protocols for forestry. Technical Memorandum prepared for the California Forestry Association, CH2M HILL, Sacramento

MCPFE (2009) Ministerial conference on the protection of forests in Europe. Available at http://www.mcpfe.org

Miner R, Perez-Garcia J (2007) The greenhouse gas and carbon profile of the global forest products industry. Forest Products J 57:80–90

Montréal Process (2005) Montréal process. Available at http://www.rinya.maff.go.jp/mpci

Nabuurs GJ, Ravindranath NH, Paustian K, Freibauer A, Hohenstein W, Makundi W (2003). LUCF Sector Good Practice Guidance (Chap. 3) Appendix 3A.1. Harvested wood products: basis for future methodological development. In: IPCC (2003) Penman J, Gytarsky M, Hiraishi T, Krug T, Kruger D, Pipatti R, Buendia L, Miwa K, Ngara T, Tanabe K, Wagner F, Good Practice Guidance for Land Use, land-Use Change and Forestry IPCC/IGES, Hayama, Japan

Nabuurs GJ, Masera O, Andrasko K, Benitez-Ponce P, Boer R, Dutschke M, Elsiddig E, Ford Robertson J, Frumhoff P, Karjalainen T, Krankina O, Kurz WA, Matsumoto M, Oyhantcabal W, Ravindranath NH, Sanz Sanchez MJ, Zhang X (2007) Forestry. In: IPCC (2007) Climate Change 2007: Mitigation. Contribution of Working Group III to the Fourth Assessment Report of the Intergovernmental Panel on Climate Change [B. Metz, O.R. Davidson, P.R. Bosch, R. Dave, L.A. Meyer (eds)], Cambridge University Press, Cambridge, United Kingdom and New York, NY, USA

Papale D et al (2005) Assessment of full carbon budget of Italy: the CarbIUS project. In: Proceedings of the 4th Annual Conference on Carbon Capture and Sequestration, Alexandria, Virginia, Paper 145. Monitor and Exchange Publications and Forum, Washington DC, 2–5 May 2005

Pettenella D, Ciccarese L (2009) Stock e flussi nel sistema forestale. Un tentativo di lettura incrociata dei dati italiani. Sherwood, Foreste ed Alberi Oggi 154:5–13

Pingoud K, Skog K, Martino DL, Tonosaki M, Xiaoquan Z (2006) Harvested Wood Products (Chap. 12). In: Eggleston HS, Buendia L, Miwa K, Ngara T, Tanabe K (eds) IPCC guidelines for national greenhouse gas inventories, vol 4: Agriculture, forestry and other land use. The Institute for Global Environmental Strategies (IGES) for the Intergovernmental Panel on Climate Change (IPCC), Hayama

Schlamadinger B, Marland G (1996) The role of forest and bioenergy strategies in the global carbon cycle. Biomass Bioenergy 10:275–300

Schroeder TA, Cohen WB, Yang Z (2007) Patterns of forest regrowth following clearcutting in western Oregon as determined from a landsat time-series. For Ecol Manage 243:259–273

Skog K, Nicholson G (1998) Carbon cycling through wood products: the role of wood and paper products in carbon sequestration. Forest Products J 48(7/8):75–83

Skog K, Kenneth E, Nicholson, Geraldine A (2000) Carbon sequestration in wood and paper products. In: Joyce LA, Birdsey R (eds) The impact of climate change on America's forests: a technical document supporting the 2000 USDA Forest Service RPA Assessment. Gen Tech Rep RMRS-GTR-59, Fort Collins, Department of Agriculture, Forest Service, Rocky Mountain Research Station. pp 79–88

Souza JCM, Roberts DA, Cochrane MA (2005) Combining spectral and spatial information to map canopy damage from selective logging and forest fires. Remote Sens Environ 98:329–343

Taverna R, Hofer P, Werner F, Kaufmann E, Thürig E (2007) The CO_2 effects of the Swiss forestry and timber industry. Scenarios of future potential for climate-change mitigation. Environmental Studies No. 0739. Federal Office for the Environment, Bern, 102 p

UNFCCC (2003) Estimation, reporting and accounting of harvested wood products. United Nations Framework Convention on Climate Change, Geneva

UNFCCC (2007) National inventory submissions 2007. United Nations Framework Convention on Climate Change. Available on-line http://unfccc.int/national_reports/items/1408.php. Last accessed Apr 2010

Winjum JK, Brown S, Schlamadinger B (1998) Forest harvests and wood products: sources and sinks of atmospheric carbon dioxide. Forest Sci 44:272–284

Wulder MA, Hana T, White JC, Sweda T, Tsuzuki H (2007) Integrating profiling LIDAR with Landsat data for regional boreal forest canopy attribute estimation and change characterization. Remote Sens Environ 110:123–137

Part III
Croplands, Grasslands and Natural Ecosystems

Chapter 8
Soil Carbon Stocks and Fluxes

Tommaso Chiti, Costantino Sirca, Mirco Rodeghiero, Donatella Spano and Riccardo Valentini

Abstract The aim of this chapter was to quantify the soil organic carbon (SOC) stock in the top 30 cm of mineral soil for the whole Italian territory, according to the different land use types of the Intergovernmental Panel on Climate Change (IPCC) cropland category (arable land, agroforestry, vineyards, olive groves, orchards and rice fields), as a basis for future land use scenarios and to address mitigation policy at country level. Besides, two independent studies addressing the current status and the future trends of SOC for the whole cropland category at regional level were reported. The subdivision of the cropland category into classes is functional to assess the impacts on the SOC stock due to land use changes from and to agricultural uses, providing the starting or ending point scenario. The differences emphasized for the soils of a subcategory under the different types of climate can be possibly used for future-oriented agricultural practices. The comparison of the total mean values of the different cropland subcategories shows significant differences in the SOC stock. Considering the year 2000 and applying to each subcategory area the specific average SOC stock value found in this study, the total amount of C stored in the upper 30 cm

T. Chiti (✉) · R. Valentini
Department for Innovation in Biological, Agro-food and Forest System (DIBAF), University of Tuscia, via San C. de Lellis snc, 01100 Viterbo, Italy
e-mail: tommaso.chiti@unitus.it

T. Chiti · R. Valentini
Impacts on Agriculture, Forest and Natural Ecosystem Division (IAFENT), Euro-Mediterranean Center on Climate Changes (CMCC), Viterbo, Italy

C. Sirca · D. Spano
Department of Science for Nature and Environmental Resources (DIPNET), University of Sassari, via E. de Nicola 9, 07100 Sassari, Italy

C. Sirca · D. Spano
Impacts on Agriculture, Forest and Natural Ecosystem Division (IAFENT), Euro-Mediterranean Center on Climate Changes (CMCC), Sassari, Italy

M. Rodeghiero
Research and Innovation Centre—Fondazione Edmund Mach, via Mach 1, 38010 San Michele all'adige, TN, Italy

R. Valentini and F. Miglietta (eds.), *The Greenhouse Gas Balance of Italy*, Environmental Science and Engineering, DOI 10.1007/978-3-642-32424-6_8

of the whole cropland category results to be 516.3 ± 156.9 Tg C. This amount represents about 17 % of the total SOC estimated for the top 50 cm of soils of total surface of Italy, which reports about 2,900 Tg C, thus indicating the importance to preserve the large amount of SOC stored in cropland category. In conclusion, given the few estimates available at European level, repeated SOC inventories aimed to define the SOC content in cropland soils are important for future stock change evaluation.

8.1 Introduction

Cropland is among the major land use in Europe, occupying around 126.5 Mha in the EU25 plus Norway and Switzerland (Janssens et al. 2005), hence changes in the size of the cropland soil carbon pool could have significant impacts on the European carbon budget (Janssens et al. 2003). As a matter of fact, under the Kyoto Protocol (UNFCCC 1998), carbon (C) sequestration in agricultural soils is accountable under Article 3.4, being the management of cropland among those activities that a Country signatory of the Kyoto Protocol may elect for meeting its emission reduction target. Soil organic carbon (SOC) stock and dynamics in cropland soils are also important for the quantification of the C emissions/removals involved in land use change processes, such as afforestation, reforestation and deforestation activities, which are to be accounted under Article 3.3 of the Kyoto Protocol.

Although the importance of the sector is well recognized, the information published so far is scarce, fragmented, not homogeneous and limited to spotted studies at local scale, that is also the reason why Italy has not decided to elect cropland management among the additional activities of article 3.4 (Tedeschi and Lumicisi 2006).

In this purpose, the general aim of this chapter is to provide the most updated information on the SOC stock stored at national level for the IPCC cropland category. Specific aims are:

- To determine the amount of SOC stored in the different cropland land use types at national level;
- To provide information on the agricultural practices able to preserve or increase the SOC stock of the cropland categories;
- To determine the future trends of the SOC stored in the cropland soils trough the use of a SOC model, considering as a case study the Sardinia region;
- To asses the SOC fluxes along two specific case studies from Trentino Alto Adige region.

8.2 Croplands SOC Stocks Under Different Types of Land Use and Climate

In order to provide Italy with the most updated information on the SOC stock for the different land use types comprised within the IPCC cropland category (e.g. arable land, agroforestry, vineyards, olive groves, orchards, and rice fields), according to the different climatic zones and landscapes of the country, a pedological database derived from different sources was used. A national project started in 2007 within the framework of the Multi-Scale Soil Information System (MEUSIS) European project and denominated "SIAS", Development of Soil Indicators in Italy, was the main source of information for the SOC stock (http://eusoils.jrc.ec.europa.eu/projects/Meusis/italy.html). The SIAS database reports the elaborations and the metadata for soil organic C stock in agreement with the IPCC standards (IPCC 2003). Following the European Directive denominated "INfrastructure for SPatial InfoRmation for Europe" (INSPIRE, COM 516/2004), the SIAS data refers to the INSPIRE grid having cells of 1 by 1 km. The value for each cell is an average value of the C stock of the upper 30 cm of mineral soil, weighed on the real surface of the cell covered by soil. For the country areas not covered by the SIAS project, an INSPIRE grid was created according to the guidelines available at the European Soil Bureau website (http://eusoils.jrc.ec.europa.eu). The analytical data to calculate the soil carbon stock of the representative profiles for each INSPIRE cell were obtained from regional reports. The data coverage of the Italian country are reported in Fig. 8.1a.

The climatic subdivision of the Italian territory was done following the "Georeferenced Soil Database for Europe: Manual of Procedures Version 1.1. Eur 18092 EN" (Finke et al. 1998). Accordingly, the Italian territory was classified in seven areas with a different type of climate (Fig. 8.1b). The CORINE Land Cover 2000 (Sinanet 2009), the climatic database, the soil C stock databases and

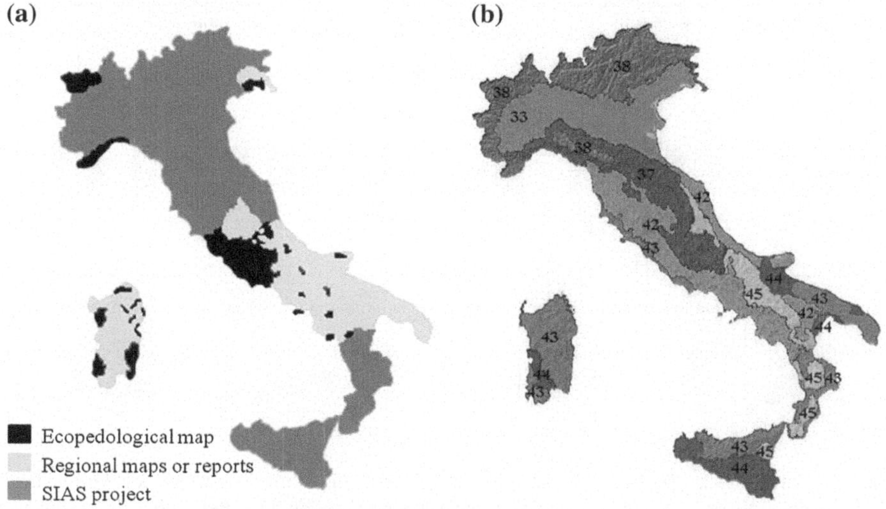

Fig. 8.1 On the *left side* **a**, coverage distribution of the different data sources used in this study. On the *right side* **b**, climatic subdivision of the Italian territory. Letters indicate the different climate types: *33* Temperate sub oceanic, *37* Temperate subcontinental to temperate continental partly mountainous, *38* Warm temperate sub continental, *42* Mediterranean oceanic to Mediterranean sub oceanic partly mountainous, *43* Mediterranean sub continental to Mediterranean continental, *44* Mediterranean to sub tropical, *45* Mediterranean mountainous

the Ecopedological map, were overlayed using the ESRI ARC/GIS software thus enabling the climate and soil-land-use combinations to be defined. The mean SOC stock value was consequently calculated for each cropland land use type combination. An analysis of variance was made (ANOVA), to statistically compare the SOC stock means, clustered according to the different climate regions, land use types and landscapes.

The land use types "arable land" and "agroforestry" are the only being present in all of the different climatic regions of Italy. The arable land SOC stock does not show substantial differences between the different climate types with an amount of SOC comparable to those of the agroforestry subcategory with a SOC stock varying from 40.1 $\pm$ 2.3 Mg C ha^{-1} in the Mediterranean subtropical region to 69.8 $\pm$ 25.3 Mg C ha^{-1} in the temperate mountain one (Fig. 8.2).

Vineyards' SOC stock varies from 39.2 $\pm$ 10.0 Mg C ha^{-1} on mountains relief of the warm Mediterranean sub continental type of climate, to

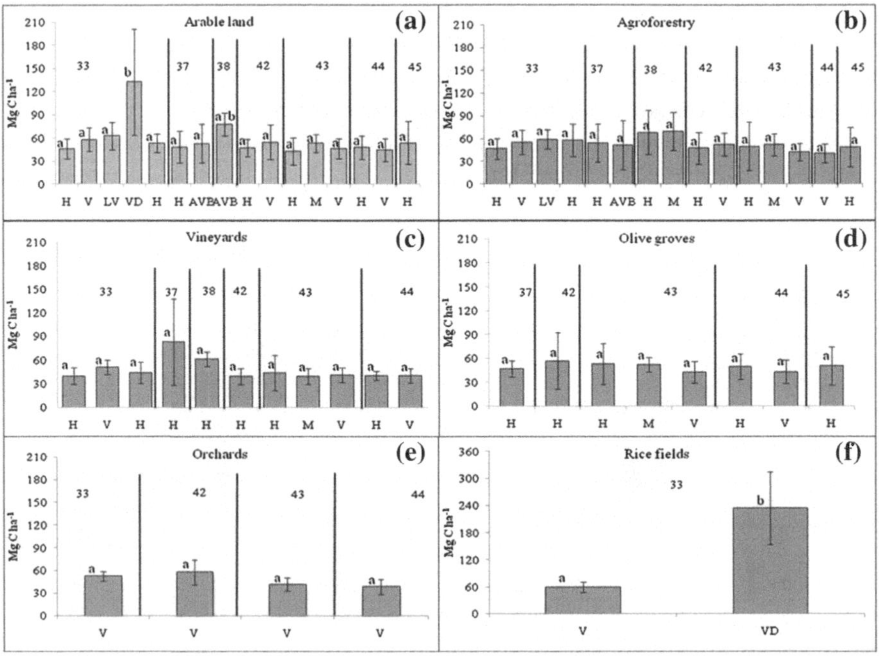

Fig. 8.2 SOC stock (2000) in the top 30 cm mineral soil of the subcategories Arable land, Agroforestry, Vineyards, Olive groves, Orchards and Rice fields under the different type of climates: temperate sub oceanic (*33*), temperate subcontinental to temperate continental partly mountainous (*37*), warm temperate sub continental (*38*), Mediterranean oceanic to Mediterranean sub oceanic partly mountainous (*42*), Mediterranean sub continental to Mediterranean continental (*43*), Mediterranean to sub tropical (*44*), Mediterranean mountainous (*45*). *H* hills; *V* valley; *LV* low valley; *VD* valley depression; *AVB* alluvial valley bottom; *AV* alluvial valley; *M* mountains. *Bars* represent the standard deviation. *Different letters* indicate significant difference ($p < 0.05$)

82.8 ± 54.9 Mg C ha^{-1} on the hills of the warm temperate sub continental type of climate (Fig. 8.2). The olive groves subcategory is obviously present only in the most typically Mediterranean type of climates with no significant statistical differences ($p < 0.05$), with SOC stock values varying from 42.1 ± 13.3 Mg C ha^{-1} in the lowland of the Mediterranean sub-continental to continental type of climate, to 56.0 ± 35.0 Mg C ha^{-1} in oceanic type of climate and in the Mediterranean sub-oceanic. The orchards' SOC stock varies from 38.1 ± 9.9 to 57.8 ± 16.4 Mg C ha^{-1} while the rice category is varying from 60.1 ± 11.1 to 234.4 ± 80.3 Mg C ha^{-1} (Fig. 8.2).

8.3 Trends in Cropland SOC Stocks Variations

The subdivision in different land uses for the cropland category in 2000 is functional to assess the impacts on the SOC stock due to land use changes from and to agricultural uses, providing the starting or ending point scenario. In the literature few works distinguish the SOC stock under the different cropland subcategories. The comparison with the values reported for Spain by Murillo (2001) indicate the presence of similar stocks in the top 30 cm of some subcategories such as for arable lands (from 50.8 ± 33.7 to 57.6 ± 36.3 Mg C ha^{-1}), for vineyards (42.5 ± 28.9 Mg C ha^{-1}) and for olive groves (39.9 ± 28.3 Mg C ha^{-1}), respectively. Similar stocks were reported also by Martin et al. (2010) for France: about 49.3 ± 26.2 Mg C ha^{-1} for agricultural soils, about 48.1 ± 24.3 Mg C ha^{-1} for orchards and 39.4 ± 26.5 Mg C ha^{-1} for vineyards. Contrasting the total mean value of the whole cropland category estimated in this study, 53.2 ± 22.2 Mg C ha^{-1} obtained by taking into account each subcategory value and its relative surface in 2000 (ISTAT 2010), with other estimates at national level indicate similar SOC stocks. For Belgium, Sleutel et al. (2006) set the average SOC stock for cropland soils (0–30 cm) at 50–61 Mg C ha^{-1}, while Smith et al. (2001) suggested value of 53 Mg C ha^{-1} in an average estimate of all European croplands. On the other hand, for Kyoto purposes the required data are not the SOC stocks but the dynamics of the soil carbon pool. In particular it is required the average carbon flux in the base year (1990), that has consequently to be compared with the average flux during the period 2008–2012.

Looking at the SOC dynamics of cropland soils at Italian level, the SOC stock in the 1990 for the whole cropland category can be taken from the work of Fantappiè et al. (2010) that report the data of the upper 50 cm of mineral soil derived from different surveys at country level from 1979 to 1998 obtained from the National soil database (Costantini et al. 2007). Accordingly, a derived value of 54.8 ± 12.2 Mg C ha^{-1} for the upper 30 cm of soil in the 1990 suggests a variation of 0.16 Mg C ha^{-1} y^{-1} when compared with the stock estimated in this study for the year 2000, 53.2 ± 22.2 Mg C ha^{-1}. A small decrease in the SOC stock from 1990 to 2000 would be in agreement with the loss rates of 0.2 Mg C ha^{-1} y^{-1} reported for Italy by Janssens et al. (2005), calculated by multiplying country specific C sequestration rates, estimated by the CESAR model (Vleeshouwers and Verhagen 2002), with the average surface area reported by Mucher (2000) and FAO (http://www.fao.org/waicent/portal/statisticsen.asp). To our knowledge, there are only two large-scale (national) and long-term inventories of C in agricultural soils that can be used to compare these SOC loss rates. In a study by Sleutel et al. (2006), repeated sampling of cropland soils in Belgium (210,000 samples taken between 1989 and 1999) indicate an average annual soil C loss of 0.8 Mg ha^{-1} y^{-1}, while Dersch and Boehm (1997) in a large-scale inventory study estimate for Austrian croplands, a net C loss of about 0.2 Mg ha^{-1} y^{-1}. Also at European level, a net loss of SOC is observed for cropland soils and can be mainly related to the harvesting that reduces C returns to soil, and to agricultural practices such as tillage (Janssens et al. 2005).

8.4 Increasing the Size of the Soc Stock for the Cropland Category

To increase the size of the cropland SOC pool, and at the same time reducing the emission from soil, it is important to take into account the different mitigation options. In fact, trying to increase the size of the soil C stock means increasing the C input, decreasing the output or a combination of the two through improved management. Carbon sequestration can also occur through a reduction in soil disturbance because more C is lost from tilled soils than from soils that are less disturbed. Measures for reducing soil disturbance include reduced or zero tillage systems, set-aside of lands and the growth of perennial crops. Measures for increasing soil C inputs include the preferential use of animal manure, crop residues, sewage sludge, improved rotations with higher carbon inputs to the soil, and in some cases fertilisation/irrigation management to increase productivity. According to Freibauer et al. (2004) and to Smith et al. (2000a, b), the impact of such practices can be estimated: for example permanent set-aside or zero tillage might result in a maximum C sequestration potential of 0.4 Mg C ha^{-1} y^{-1}, the use of perennial crops or deep rooting crops in 0.6 Mg C ha^{-1} y^{-1}, while changing from conventional to organic farming result in a sequestration potential of 0.5 Mg C ha^{-1} y^{-1}. A practical example for Italy can be done considering the 2007, the more recent year for which ISTAT report the areas occupied by the different cropland subcategories (ISTAT 2010). The average SOC stock in 2007 can be obtained by applying the annual SOC net loss rate of 0.16 Mg C ha^{-1} y^{-1} estimated in this study to the average SOC stock found in 2000 for each subcategory and multiplying the resulting stocks for the area each subcategory occupy. This calculation results in a total SOC stock for the whole cropland category of 489.7 $\pm$ 148.2 Tg C (Fig. 8.3). Consequently some future scenarios can be hypothesised (e.g. for 2020) assuming no further variations in the area occupied by each subcategory.

In 2007, the ISTAT report that about half of the surface occupied by arable lands (3.48 Mha) is currently ploughed. About 60 % of this area is ploughed

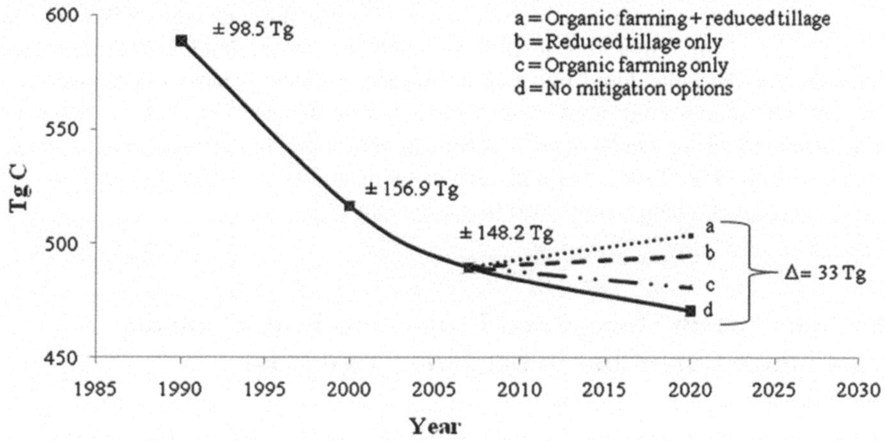

Fig. 8.3 Simulation of the impact of different mitigation options on the total SOC stock of the whole cropland category from 2007 to 2020. Numbers indicates the standard deviation of the SOC stock measured in the different years

between 20 and 30 cm, while about 10 % is ploughed to more than 40 cm. Assuming a decrease of the ploughing depth to 20 cm for the 70 % of the ploughed arable land area (2.43 Mha), and applying the corresponding C sequestration factor for reduced tillage, the amount of C sequestered in 2020 would be 10.4 Tg C, corresponding to 0.8 Tg C y^{-1}. Making a comparison with the 2007 SOC stock and considering that the other subcategories continue to lose C at the same rate, the increase in 2020 for the whole cropland category is 5.2 Tg C (Fig. 8.3). The improvement of organic farming is another interesting option to increase SOC stock in croplands. In fact, in Italy the area occupied by organic farming represents only 5 % of the total agricultural surface. It is worth nothing that, at European level, the size of organic farming soil is even less, about 2 % of the total agricultural area (Rounsevell et al. 2002), and Italy alone has 27 % of European Union organic land (EC 2001). A 2 % increase of the area occupied in 2007 by organic farming (0.5 Mha) is leading to an increase of about 0.8 Tg C y^{-1} until 2020, not enough to change the negative trend of the whole cropland category. In fact, despite the similarity in the accumulation rate with reduced tillage option, the loss of C from the other subcategories has to be considered, bringing to an estimate for the total SOC stock in 2020 of 480.9 Tg C, 8.8 Tg C less than the amount in 2007 (Fig. 8.3). Combining the two mitigation options change the negative trend of SOC from the cropland category, leading to a total SOC stock in 2020 of 503.6 Tg C, 13.9 Tg C more than in 2007. More important, in 2020 the increase resulting from combining reduced till age and organic farming mitigation options will be about 33 Tg C more than the amount hypothesisable if no mitigation options are planned, about 470.5 Tg C (Fig. 8.3). All these hypothetical increments indicate the difficulty in a sensible size increase of the cropland SOC pool unless a national specific policy aimed to improve the agricultural practices is promoted. In fact, it is not realistic to hypothesize that these changes will occur in a short lapse of time, therefore leading to much smaller increments than those we estimated for the considered mitigation options. Besides, it has to be considered that the SOC sequestered in cropland soils is not permanent. By changing agricultural management or land-use, soil carbon is lost more rapidly than it can accumulate (Smith et al. 1996). For soil carbon sequestration to occur, the land-use or land-management change must also be permanent.

8.5 Soil Carbon Storage and Projections Under Climate Change Conditions in Sardinia Croplands

In this paragraph, a regional estimation of SOC pools in Mediterranean cropland soils is showed. RothC, a dynamic simulation model of carbon turnover in the soil (Coleman and Jenkinson 1996) was linked to a GIS soil and climate database to (i) assess the actual SOC stocks in croplands; (ii) evaluate the model performance by comparison of modeled and observed SOC data; (iii) estimate the future SOC pool variation under a climate change scenario.

The present study was performed in the Sardinia region. Sardinia is the second largest island in the Mediterranean Sea (about 24,090 km^2). The climate is Mediterranean, sub-arid type with a warm summer, mild winter, and a remarkable water deficit from May to September. Most of the rainfall (approximately 700 mm/year) occurs in fall and winter. The average annual temperature is 15 °C, while the average minimum one is 7 °C, and the average maximum temperature is 28 °C. Almost 28 % of the total island soils are represented by a mix formed by rock outcrops and Leptosols (Eutric, Dystric e Lithic), spread throughout the territory and especially on hard rock (metamorphic, intrusive, effusive, dolomitic and limestone), mostly in steep areas with irregular shape and free of tree cover or shrublands. The surface with agricultural use (including crops, sown field and orchards) is equal to the 38 % of the whole island surface (AA.VV 2008). The two main tree crops are olive groves (48,779 ha) and vineyards (24,693 ha). The remaining tree crops cover less than 12,000 ha. The herbaceous irrigated crops are the 26.1 % of the island surface. Table 8.1 reports the Land Use categories included in the present analysis.

Table 8.1 Agricultural land use categories* where the analysis was performed

Land use	Surface (ha)	Surface (%)
Permanent irrigated and irrigated arable lands	628,326	26.1
Vineyard	24,693	1.0
Olive groves	48,779	2.0
Total	701,798	29.1

**Source* Land use map of Sardinia 2008

The SOC pools were assessed using RothC (v. 26.3) model developed to estimate the turnover of carbon in soil at monthly temporal step (Coleman and Jenkinson 1996). RothC includes four soil C compartments (DPM—Decomposable Plant Material, RPM—Resistant Plant Material, BIO—Microbial Biomass, and HUM—Humified Organic Matter). It requires input data at monthly scale, related to climate (rainfall, evapotranspiration, air temperature), soil (clay content, cover), and plant residues (litterfall). With respect to the litterfall, the model is designed to run in two modes: 'forward', which permits to calculate changes in soil organic matter using available litterfall input, and 'inverse' mode, which calculates the litterfall rates from the values of SOC.

The Soil Map (for clay content) and the Land Use (LU) Map of Sardinia were processed using a GIS, and almost 380,000 polygons (homogeneous for soil characteristics and Land Use class) were detected. In each polygon covered by crops, the actual SOC (related to the top 0.24 m of mineral soil) was estimated using RothC coupled with the actual climate data (available at 250 m resolution). We used literature and our unpublished data for the input data related to plant residues. The scheme of the methodology used is reported in Fig. 8.4. The model results were compared with soil analysis values made in different Land Use typologies, and the litterfall input for each LU class was adjusted to find the best fit ('inverse' run of the model). After a validation using others independent measured SOC points, a new run of RothC was performed to obtain a SOC database of Sardinia croplands for the actual climate and under climate change conditions (A1B scenario, source: CMCC, Euro-Mediterranean Centre for Climate Change).

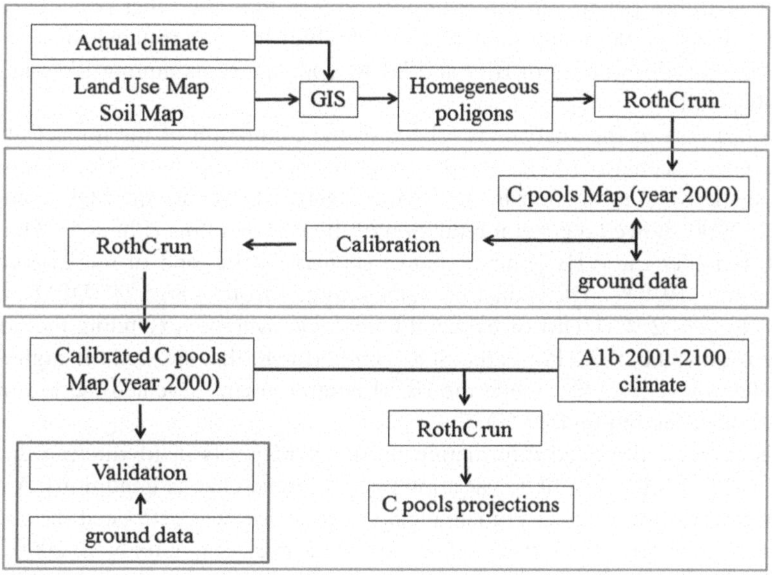

Fig. 8.4 Scheme of the methodology

Table 8.2 Total SOC and quantities estimated per unit of soil surface (ha)

Land use category	Total SOC (Mg × 1,000)	SOC (Mg ha^{-1})
Permanent irrigated and irrigated arable lands	19,852.5	31.6
Vineyard	834.8	33.8
Olive groves	1,568.5	32.2
Total	22,255.7	

Table 8.3 SOC pools for the different croplands

Land use category	DPM	RPM	BIO	HUM
	(Mg ha^{-1})			
Permanent irrigated and irrigated arable lands	0.22	4.79	0.69	25.90
Vineyard	0.20	9.30	0.63	23.68
Olive groves	0.01	8.00	0.62	23.52

DPM decomposable plant material; *RPM* resistant plant material; *BIO* microbial biomass; *HUM* humified organic matter

The total SOC for the cropland included in the analysis is calculated at about 22,256 Mg under current climate (Table 8.2). Even if differences in the total SOC values of the three Land Use classes were observed, the SOC content per ha shows similar values for the three categories.

Larger differences among Land Use categories for the SOC compartments were observed (Table 8.3). The HUM compartment, in general, is the largest SOC pool. The DPM values per ha are lower in olive groves than the other two LU categories. The RPM contents are higher (about double) in tree cropland than irrigated lands. Smaller differences in BIO and HUM compartments among LU categories were observed.

The last part of the analysis is related to the evaluation of the impact of a climate change scenario (A1B) on the SOC pools in croplands. The comparison between the SOC projection in the period 2091/2100 versus the SOC content in the year 2000 showed a general reduction of the carbon pools (Fig. 8.5). The mean SOC loss under the A1B climate change scenario at the end of the 21st century is equal to −19.1 %. The reduction values ranged from −86.5 % (DPM of olive groves) to −14.2 % (HUM of Irrigated lands). On average, excluding the DPM of olive groves value, the RPM is the SOC compartment that showed the highest carbon reduction (−31.2 %), while the HUM compartments revealed the lower average carbon reduction (−15.2 %).

This exercise provided information on the SOC pools in Mediterranean croplands, and projections of SOC variation under climate change conditions. A reduction in the SOC pool under climate change conditions was estimated in Sardinian croplands, as reported by other authors for other areas (e.g. Cox et al. 2000; Jones et al. 2003). The inverse model run also allowed the estimate and comparison with literature values related to the litterfall rates.

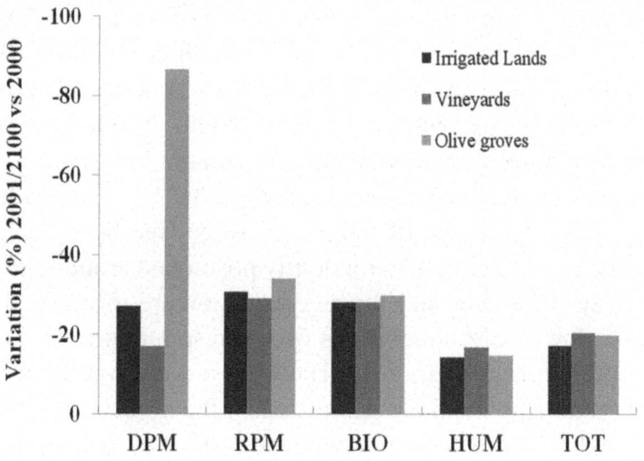

Fig. 8.5 Percentage of variation in the organic carbon pools with the A1b climatic scenario (2091/2100 mean SOC data vs. 2000)

8.6 Regional Case Study on SOC Fluxes from Trentino

The site "Monte Bondone" (see Sect. 2.2.13) is equipped with an Eddy Covariance measurement tower. The meadow is mowed once a year in mid July. Soil CO_2 fluxes were measured at twenty sampling points with a closed dynamic system (LI-COR 6400, soil chamber LI 6400-09; LI-COR Inc., Lincoln, NE, USA) every 15–20 days. The soil chamber was used with plastic collars inserted for about 1.5 cm into the soil and fixed with iron legs to prevent the collar from moving when the chamber was placed on it. The collars were systematically positioned around the eddy flux tower following a cross pattern. The vegetation inside the collars was cut 1 day prior to measurement. Soil temperature at a depth of 5 cm was measured with the LI-COR 6400 temperature probe. Soil water content was recorded on three points around each soil collar with a ThetaProbe ML2x (Delta-T Devices Ltd, Cambridge, UK). Moreover the soil temperature at a depth of 5 cm was recorded every 5 min with StowAway TidbiT loggers (Onset Comp. Corp., Bourne, MA, USA). An exponential model (Bahn et al. 2008; Rodeghiero and Cescatti 2005) was fitted to the temperature (T_S), soil respiration (R_S) data:

$$R_s = R_{10} * \exp(E_0 * (1/56.02 - (1/T_s + 46.02))) \quad (8.1)$$

where R_{10} is the soil respiration at 10 °C, E_0 is the activation energy (Lloyd and Taylor 1994; Eq. 8.1). The total soil respiration in a given year was calculated by summing the fluxes estimated by the model from continuous temperature data.

In this meadow ecosystem temperature resulted to be the main determinant of soil respiration [the variance explained by the fitting model being 0.77 (Eq. 8.1; Fig. 8.6a)] whereas no evident drought stresses during the growing season were detected. A reduction of the soil CO_2 flux rates after mowing was observed and was probably related to a decrease in photosynthetic activity which caused a consequent decrease in autotrophic respiration (data not shown). As a consequence, the higher level of irradiance lowered the soil water content and possibly diminished the heterotrophic respiration as well. With an annual soil CO_2 efflux of 17.4 Mg C ha^{-1} y^{-1} (year 2003; Bahn et al. 2008) this site resulted to have among the highest level of CO_2 rates among a series of thirteen European grasslands and meadow ecosystems (Bahn et al. 2008). The high 2003 CO_2 emissions were due to the high average annual soil temperature being 9.7 °C at 5 cm depth. In 2004 there was a much lower efflux (11.0 Mg C ha^{-1} y^{-1}) and a lower soil temperature (6.8 °C).

Also in the site of Valle dell'Adige (see Sect. 2.2.25) soil CO_2 efflux was measured on twenty collars randomly positioned around the eddy flux tower. The measurement techniques and fluxes calculation are the same as those reported above for the meadow site. Annual totals of soil respiration were evaluated by applying the above reported model (Eq. 8.1). The total soil respiration in a given year was calculated by summing up the fluxes estimated by the model from continuous temperature data.

The temperature dependency of soil respiration was even higher than in the meadow, being the variance accounted for by the model 0.83 (Fig. 8.6b).

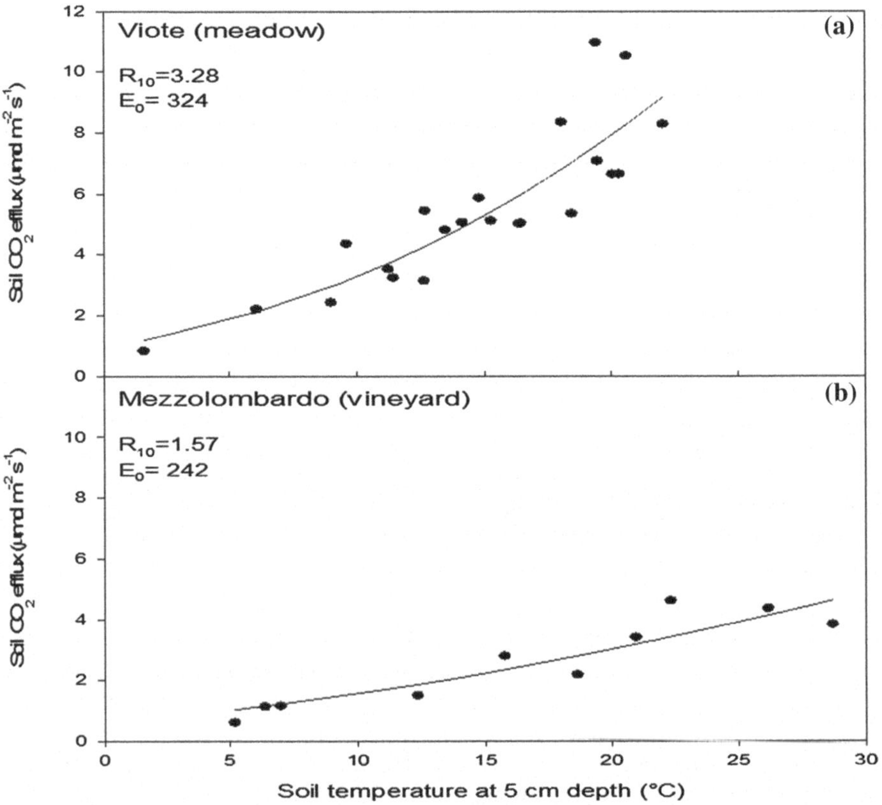

Fig. 8.6 Temperature dependency of soil CO_2 efflux for two sites located in Trentino region according to the model described in Eq. 8.1. The meadow (**a**) has higher values both of respiration at 10 °C (R_{10}) and activation energy (E_0) compared to the vineyard (**b**). See text for further details

However, the respiration rates of the vineyard were lower compared to the meadow. The average annual soil temperature at 5 cm depth was almost the same in 2008 and 2009 (14.6 °C) leading to similar annual soil CO_2 fluxes: 8.6 and 8.8 Mg C ha^{-1} y^{-1} respectively. In 2010 there was a slight decrease of soil temperature up to 12.5 °C with a soil CO_2 efflux of 8.6 Mg C ha^{-1} y^{-1}.

8.7 Conclusions

Looking at the SOC dynamics for croplands at the national level, recent studies indicate that the SOC pool is near to the equilibrium with an average loss of about 0.2–0.5 Mg C ha^{-1} y^{-1} (Gardi and Sconosciuto 2007; Janssens et al. 2005; Lugato et al. 2010; Morari et al. 2006). This loss of SOC is possibly attributable to the intensification of agricultural practices in soils cultivated for thousands of years (Gardi and Sconosciuto 2007; Morari et al. 2006) and could be reduced by taking into account some of the different mitigation options such as reduced tillage, improved management or the use of animal manure (Freibauer et al. 2004; Smith et al. 2000a, b; Triberti et al. 2008). Also, the modelling exercise performed in Sardinia indicate as the SOC pool for cropland in the Mediterranean area is decreasing as a consequence of climate change and the continuous exploitation of soil for agricultural purpose. The losses could even increase as a consequence of the increase in temperature, the soil CO_2 efflux being very sensitive to temperature rather than other parameters, as suggested by the case study from Trentino Alto Adige.

In conclusion, the assessment of the average SOC stock of the different cropland land uses, landscape position and climate regions could notably help when assessing the impact of different agricultural practices and future stock change evaluation.

References

AA.VV (2008) Carta di Uso del Suolo della Regione Sardegna. Produced by the Regional Government of Sardinia

Bahn M, Rodeghiero M, Anderson-Dunn M, Dore S, Gimeno C, Drösler M, Williams M, Ammann C, Berninger F, Flechard C, Jones S, Balzarolo M, Kumar S, Newesely C, Priwitzer T, Raschi A, Siegwolf R, Susiluoto S, Tenhunen J, Wohlfhart G, Cernusca A (2008) Soil respiration in European grasslands in relation to climate and assimilate supply. Ecosystems 11:1352–1367

Coleman K, Jenkinson DS (1996) RothC-26.3. A model for the turnover of carbon in soil. In: Powlson DS, Smith P, Smith JU (eds) Evaluation of soil organic matter models using existing, long-term datasets. NATO ASI Series I, vol 38, pp 237–246. Springer, Berlin

Costantini EAC, Barbetti R, L'Abate G (2007) Soils of Italy: status, problems and solutions. In: Zdruli P, Trisorio Liuzzi G (eds) Status of mediterranean soil resources: actions needed to support their sustainable use. Mediterranean conference proceedings, Tunis, Tunisia, IAM Bari (Italy), pp 165–186

Cox PM, Betts RA, Jones CD, Spall SA, Totterdell IJ (2000) Acceleration of global warming due to carbon-cycle feedbacks in a coupled climate model. Nature 408:184–187

Dersch G, Boehm K (1997) Bodenschutz in Österreich. In: Klaghofer E, Loechl A et al (eds) Blum WEH. Bundesamt und Forschungszentrum fuer Landwirtschaft, Österreich (in German), pp 411–432

European Commission (EC) (2001) Agriculture in the European union: statistical and economic information 2000. http://europa.eu.int/comm/agriculture/agrista/2000/table-en/index.htm

Fantappiè M, L'Abate G, Costantini EAC (2010) Factors influencing soil organic carbon stock variations in Italy during the last three decades. In: Zdruli et al (eds) Land degradation and desertification: assessment, mitigation and remediation, pp 435–465. doi:10.1007/978-90-481-8657-0_34

Finke P, Hartwich R, Dudal R, Ibanez J, Jamagne M, King D, Montanarella L, Yassoglu N (1998) Georeferenced soil database for Europe. Manual of procedures. Version 1.0 . ESB-JRCSAI. European Commission. EUR 18092 EN, p 184

Freibauer A, Rounsevell MDA, Smith P, Verhagen J (2004) Carbon sequestration in the agricultural soils of Europe. Geoderma 122:1–23

Gardi C, Sconosciuto F (2007) Evaluation of carbon stock variation in Northern Italian soils over the last 70 years. Sustain Sci 2:237–243

IPCC (2003) Penman J, Gytarsky M, Hiraishi T, Krug T, Kruger D, Pipatti R, Buendia L, Miwa K, Ngara T, Tanabe K, Wagner F, Good Practice Guidance for Land Use, land-Use Change and Forestry IPCC/IGES, Hayama, Japan

ISTAT (2010) Agricoltura e ambiente. L'indagine 2007 sulla struttura e le produzioni delle aziende agricole. Informazioni n. 2 Roma 2010, p 71

Janssens IA, Freibauer A, Schlamadinger B, Ceulemans R, Ciais P, Dolman AJ, Heimann M, Nabuurs GJ, Smith P, Valentini R, Schulze ED (2005) The carbon budget of terrestrial ecosystems at country-scale-a European case study. Biogeosciences 2:15–26

Janssens IA, Freibauer A, Ciais P, Smith P, Nabuurs GJ, Folberth G, Schlamadinger B, Hutjes RWA, Ceulemans R, Schulze E-D, Valentini R, Dolman AJ (2003) Europe's terrestrial biosphere absorbs 7 to 12 % of European anthropogenic CO_2 emissions. Science 300:1538–1542

Jones CD, Cox PM, Essery RLH, Roberts DL, Woodage MJ (2003) Strong carbon cycle feedbacks in a climate model with interactive CO_2 and sulphate aerosols. Geophys Res Lett 30:1479

Lloyd J, Taylor JA (1994) On the temperature dependence of soil respiration. Funct Ecol 8:315–323

Lugato E, Zuliani M, Alberti G, Vedove G, Gioli B, Miglietta F, Peressotti A (2010) Application of DNDC biogeochemistry model to estimate greenhouse gas emission from Italian agricultural areas at high spatial resolution. Agr Ecosyst Environ 139:546–556

Martin MP, Watterbach M, Smith P, Meersmans J, Jolivet C, Boulonne L, Arrouays D (2010) Spatial distribution of soil organic carbon stocks in France. Biogeosci Discuss 7:8409–8443

Morari F, Lugato E, Berti A, Giardini L (2006) Long term effects of recommended management practices on soil carbon changes and sequestration in north-eastern Italy. Soil Use Manag 22:71–81

Murillo JCR (2001) Organic carbon content under different types of land use and soil in peninsular Spain. Biol Fert Soils 33:53–61

Mucher CA (2000) PELCOM (Pan-European land use and land cover monitoring, ALTERRA, Wageningen). Available online at www.geo-informatie.nl/cgi/projects/eu/pelcom/index.htm

Rodeghiero M, Cescatti A (2005) Main determinants of forest soil respiration along an elevation/temperature gradient in the Italian Alps. Glob Change Biol 11:1024–1041

Rounsevell MDA, Audsley E, Mortimer D (2002) The impact of the Common Agricultural Policy on land use in Europe, in Land Cover and Land Use. In: Verheye WH (ed) Encyclopedia of life support systems (EOLSS). Developed Under the Auspices of the UNESCO, EOLSS Publishers, Oxford

Sinanet (2009) Rete del sistema Informativo Nazionale Ambientale. http://www.clc2000.sinanet.apat.it. Cited 13 Aug 2009

Sleutel S, De Neve S, Singier B, Hofman G (2006) Organic C levels in intesively managed arable soils—long term regional trends and characterization of fractions. Soil Use Manage 22:188–196

Smith P, Powlson DS, Glendining MJ (1996) Establishing a European soil organic matter network (SOMNET). In: Powlson DS, Smith P, Smith JU (eds) Evaluation of soil organic matter models using existing, long-term datasets. NATO ASI Series I, vol 38. Springer, Berlin, pp 81–98

Smith WN, Desjardins RL, Patty E (2000a) The net flux of carbon from agricultural soils in Canada 1970–2010. Glob Change Biol 6:557–568

Smith P, Powlson DS, Smith JU, Falloon P, Coleman K (2000b) Meeting Europe's climate change commitments: quantitative estimates of the potential for carbon mitigation by agriculture. Glob Change Biol 6:525–539

Smith P, Smith JU, Powlson DS (2001) Soil organic matter network (SOMNET): 2001 model and experimental metadata. GCTE Report 7, 2nd edn, GCTE Focus 3, Wallingford, Oxon, 223 p

Tedeschi V, Lumicisi A (2006) L'attuazione del Protocollo di Kyoto nel settore forestale: il punto sulla situazione attuale e le prospettive future. Forest@ 3:3–5

Triberti L, Nastri A, Giordani G, Comellini F, Baldoni G, Toderi G (2008) Can mineral and organic fertilization help sequestrate carbon dioxide in cropland? Eur J Agron 29:13–20

UNFCCC (1998) Report of the conference of the parties on its third session, held at Kyoto from 1 to 11 Dec 1997. Addendum document FCCC/CP/1997/7/Add1. Available on the Internet http://www.unfccc.de

Vleeshouwers LM, Verhagen A (2002) Carbon emission and sequestration by agricultural land use: a model study for Europe. Glob Change Biol 8:519–530

Chapter 9
N_2O Emission Factors for Italian Crops

Simona Castaldi, Giorgio Alberti, Teresa Bertolini, Annachiara Forte, Franco Miglietta, Riccardo Valentini and Angelo Fierro

Abstract Nitrous oxide (N_2O) emissions from several Italian croplands along a latitudinal gradient were analyzed and the fertilizer induced emission (FIE) factor, for each single fertilization event, was calculated. Data show that the average emission factor was between 0.7 and 0.3 %, hence much lower than the IPCC EF used for temperate croplands. The relationship between N_2O production and applied N fertilization rate was exponential and not linear, although the rate of exponential increase was lower than previously reported. Maximum N_2O emission rate was correlated with magnitude of the total FIE, whereas it was inversely

S. Castaldi (✉) · T. Bertolini
Department of Environmental, Biological and Pharmaceutical Sciences and Technologies, Second University of Naples, Caserta, Italy
e-mail: simona.castaldi@unina2.it

S. Castaldi · R. Valentini
Impacts on Agriculture, Forest and Natural Ecosystem Division (IAFENT), Euro-Mediterranean Center on Climate Changes (CMCC), Viterbo, Italy

G. Alberti
Department of Agricultural and Environmental Sciences, University of Udine, Udine, Italy

A. Forte · A. Fierro
Department of Structural and Functional Biology, University of Naples 'Federico II', Naples, Italy

F. Miglietta
Institute of Biometeorology (IBIMET), National Research Council (CNR), via Caproni 8, 50145 Florence, Italy

F. Miglietta
FoxLab (Forest and Wood) E. Mach Foundation—Iasma, via E. Mach 1, 38010 S. Michele sll'Adige, TN, Italy

R. Valentini
Department for Innovation in Biological, Agro-food and Forest Systems (DIBAF), University of Tuscia, via S. Camillo de Lellis, 01100 Viterbo, Italy

R. Valentini and F. Miglietta (eds.), *The Greenhouse Gas Balance of Italy*, Environmental Science and Engineering, DOI 10.1007/978-3-642-32424-6_9

related to the length of FIE, which varied from a minimum of 8–56 days. Overall data suggest that the internationally applied emission factors for temperate crops, which are empirically derived from sites with cooler and wetter climates than the Mediterranean, would overestimate N_2O emissions for Italian crops, in particular those developing between spring and summer.

9.1 Role of Agricultural Ecosystem in the Total N_2O Atmospheric Budget

Nitrous oxide (N_2O) is the third most important greenhouse gas and plays a key role in the destruction of the ozone layer through the formation of NO_x in the stratosphere (Ravishankara et al. 2009). Soils are the main source of N_2O emissions contributing to about 53 % of the total source strength (Denman et al. 2007). About 60 % of these emissions occur in the Northern Hemisphere. In temperate areas, agricultural ecosystems are the primary source of N_2O due to the elevated input or reactive nitrogen (Nr) into the environment, mostly associated with fertilization with mineral N and animal manure, N derived from N_2-fixation (legumes and N-fixing microbes) and N from enhanced soil mineralization. In fact, N_2O atmospheric concentration has increased constantly in the last centuries (270 ppb before the Industrial Revolution to 320 ppb today) (Denman et al. 2007) concomitantly with the increase of Nr (Galloway et al. 2008). Indirect emissions of N_2O due to the release of Nr into the environment in the form of leaching, runoff, nitrogen oxides (NO_x) and NH_3 volatilization also play an important role in regional and continental budgets and are directly related to agricultural management (IPCC 2007).

9.2 Driving Factors of N_2O Emissions

A complex interplay of factors concurs to the magnitude of direct N_2O emissions from agricultural soils. Controlling factors can be divided into proximal and distal ones. The former are those having a direct influence on microbial processes which control N_2O emissions, i.e. nitrification and denitrification. N_2O is a by-product of nitrification, which although an aerobic process, leads to consistent emissions of N_2O only when oxygen tension decreases. N_2O is also an intermediate product of denitrification, carried out by anaerobic heterotrophic bacteria. Both processes are directly limited by substrate availability (NH_4^+, NO_3^-, organic C) and are controlled by temperature (Castaldi 2000), soil pH, and presence of inhibiting molecules (Castaldi et al. 2009). Soil water content exerts a direct effect on microbial survival, as in prolonged and extremely dry conditions part of the microbial community will die or change its biological form (ex. formation of resistant spores). However, water content also controls the solubility and hence the availability of substrates for microbial activity. Moreover, water content exerts diffusion constraints on soil O_2, thus increasing the probability of occurrence of micro sites of anaerobic activity where denitrification can take place (Smith 1990) and high rates of N_2O production can be obtained. In this respect, soil texture and structure also represent a proximal controlling factor. Distal controls are ecosystem processes which have a direct effect on proximal controls, such as climate, disturbance, agricultural management, NEP.

9.3 Available Approaches to Quantify Soil N_2O Emissions

As underlined in the previous paragraph, multiple factors concur to determine the temporal and spatial dynamics and the magnitude of N_2O emissions, some exerting a clear predominant role. The main drive is undoubtedly the availability of N substrate, however, a strong interacting effect exists between N substrate and soil water content, leading to high emission rates only when the soil water content is sufficiently high to enhance substrate diffusion and lower oxygen tensions (Rees et al. 2013).

Different levels of complexity can be achieved in order to predict N_2O emissions from agricultural soils. The most complex are biogeochemical models (Century, DNDC, ORCHIDEE, etc.) which simulate all the processes involved in N transformations and the interaction with proximal and distal controlling factors. These models require an elevated number of data to parameterize, initialize and eventually validate N_2O emissions, which make the scaling up at regional or national level quite difficult.

A general and simple procedure to produce estimates of N_2O emissions from agricultural practices at national scale is proposed by the IPCC Guidelines (1997, 2006). In this technical document, the amount of direct emissions of N_2O from fertilizer addition as well as the amount of N losses via leaching or volatilization, which contribute to indirect N_2O production, are calculated by multiplying emission factors “EFs” by the rates of N applications. The environmental parameters which might influence N losses are not explicitly included in the calculations with few exceptions, where EFs might change with quality or semi-quantitative classes of environmental variables. Also, direct N_2O emissions from soils are linearly dependent from fertilizer addition rates (EF 1 % of added N for temperate agricultural ecosystems).

Although at present the IPCC approach is, in most cases, the sole approach which can be used due to lack of data for a more detailed analysis, it presents limits which must be taken into account. For example, environmental conditions vary widely between countries and within each country. In Italy, for example, climate variations from sub-humid to arid bioclimate, going from the Alps to the Southern regions, might significantly influence the magnitude of direct and indirect N_2O emissions in different regions. Also, Bouwman et al. (2002b) has shown that N_2O emissions increase exponentially with increase of N fertilizer input and that soil characteristics, fertilizer type, fertilizer application mode, climate, strongly influence the total amount of N_2O efflux associated to a fertilization event. These two important considerations, which have relevant implications for the estimated magnitude of N_2O emissions, cannot be taken into account by the simple application of the general IPCC EFs.

The main limitation to improve the available estimates comes from the uneven distribution of studies over the globe. In particular, for what concerns studies in temperate areas, the totality of them refers to countries of central and northern Europe, while data from Mediterranean areas are unavailable (Bouwman et al. 2002a). Both high temperatures and relatively low rainfall regimes could strongly affect the EFs in this region.

9.4 A Comparison of N_2O Flux Strength Obtained Using Specific EFs for Mediterranean Conditions, General IPCC EFs and Empirical Modeling

In this paragraph we present a series of available data of N_2O emissions in croplands obtained during the CarboItaly project and from other studies in Italy. The data were analyzed in order to derive an average emission factor to be compared with those available in literature (IPCC 1997, 2006) or with estimates which can be derived using simple empirical models based on published global data (Bouwamn et al. 2002b).

The data used were recalculated from studies listed in Table 9.1. Most of the sites were cultivated with maize and all of them have been continuously cultivated in the last 20 years, so that we can therefore reasonably exclude the effect of land use change on the measured background fluxes. The analyzed sites present different soil managements, fertilization rates and site characteristics. At all sites N_2O fluxes were measured using closed static PVC chambers (Hutchinson and Mosier 1981; Smith et al. 1995) coupled with GC or Photoacustic analyser. Specific technical details on experimental set up and procedures can be found in the literature cited in Table 9.1.

At each site, the total amount of N_2O emitted at each single fertilization event was quantified by linear interpolation between sampling dates following the event till the flux reached a steady value close to the pre-fertilization level. The latter was instead defined as baseline flux and the fertilizer induced emission (FIE) was then calculated as the difference between the total flux minus the baseline flux. The corresponding EF was finally computed as the ratio between FIE and the amount of N added at the considered fertilization event.

Results show that for this specific crop, with its growing season in spring-summer, hence when the combination of temperature and rain regime mostly differs from non Mediterranean climate areas, the calculated emission factors were lower than the average emission factor reported by IPCC (2006) for temperate agricultural soils (1 %) (EFs reported in Table 9.2), with the exception of

Table 9.1 Main characteristics of experimental sites analysed in this study

Region	Site	Coordinates m a.s.l	Mean air T (°C)	ID	Crop type	Irrigation	Total water input rainfall + irrigation (mm)	Soil management	Previous crop	Clay (%)	Sand (%)	pH	SOC (%)	Ref.
Campania	Torre Lama	40°37′N,	17.1	1	Maize	Sub-optimal	362 + 109	Conventional	Bare soil	32.9	47	7.4	0.7	1
		14°58′E,		2	Maize	Sub-optimal	362 + 109	Minimum tillage	Bare soil	32.9	47	7.4	0.7	1
		30 m a.s.l.		3	Maize	Sub-optimal	362 + 109	Conventional	Hairy vetch	32.9	47	7.4	0.7	1
	Borgo Cioffi	40°31′N,	15.5	4	Maize	Yes	245 + 68	Ripping, cultivation	Ryegrass/clover	52	28	7.5	2.5	2
		14°57′E, 15 m a.s.l		5	Florence fennel	Rainfed	854 + 0	Ploughing, cultivation	Maize	52	28	7.5	2.5	2
Piemonte	Turin	44°53′N,	11.9	6	Maize	Rainfed	117 + 0	Conventional tillage	Maize	7.5	36	8.1	1.0	3
		7°41′E,		7	Maize	Rainfed	118 + 0	Stalk shredder and conventional tillage	Maize	7.5	36	8.1	1.0	3
		232 m a.s.l.		8	Maize	Rainfed	257 + 0	Conventional tillage	Maize	7.5	36	8.1	1.0	3
Tuscany	Pistoia	43°56′N 10°53′E 88 m a.s.l.	14.43	9	Maize		177 + 0	Ploughing, milling	Maize	10	35	5.4	2.1	4
Friuli	Beano	46°00′N	12.9	10	Maize	Yes	1585 + 0	Tillage (0–30 cm)	Maize	14.5	24.9	7.1	1.8	5
		13°01′E		11	Maize	Yes	1585 + 0	No tillage	Maize	13.2	26.0	7.1	1.8	5
		65 m a.s.l.		12	Alfalfa	Yes	1612 + 0	Tillage (0–30 cm)	Maize	15.8	29.5	7.2	1.9	5

References 1. Fierro and Forte (2012); 2. Ranucci et al. (2011); 3. Alluvione et al. (2010); 4. Castaldi et al. (2011); 5. Rees et al. (2013)

Table 9.2 N_2O emissions and EFs calculated for each single fertilization event in the sites presented in Table 9.1

ID	Baseline (B) g $N–N_2O$/ha day	FIE g N_2O–N/ha	FIE Length days	Peak max g N/ha day	N input kg N/ha	EFs% $\frac{FIE-B}{N\ input} \times 100$	Fertilizer type
1	0.88	273.7	56	13.3	130	0.21	Urea
2	0.17	212.3	45	18.9	130	0.16	Urea
3	0.21	97.7	41	7.2	130	0.07	Green manure
4	0.22	212.9	35	31.9	65	0.33	Entec 25
4	0.22	206.9	35	32.6	193	0.11	Urea
5	0.26	213.5	29	34.9	41	0.52	NK + NPK
5	0.26	NA*	NA*	NA*	64	–	$(NH_4)_2SO_4$
7	22.75	5,392.0	16	800.0	130	4.15	Urea
7	22.75	2,937.0	13	1,100.0	260	1.13	Vetch
8	29.72	3,532.5	18	540.0	130	2.72	Urea
9	0.63	31.5	8	2.5	20	0.16	NPK 8-24-24
9	0.63	233.8	10	29.4	100	0.24	NH_4NO_3 (N 34 %)
9	0.63	488.3	10	141.5	100	0.49	NH_4NO_3 (N 26 %)
10	0.63	47.1	35	0.8	54	0.52	N 18 % P_2O_5 46 %
10	0.63	214.3	21	35.4	83	0.69	N 46 %;
10	0.63	1,738.4	37	128.7	276	0.21	N 46 %
11	0.52	72.6	35	1.1	54	0.16	N 18 % P_2O_5 46 %
11	0.52	287.3	21	52.6	83	0.07	N 46 %;
11	0.52	2,374.1	37	185.0	276	0.33	N 46 %
12	0.63	129.0	0	2.6	0	–	No fertilizations
12	0.63	43.3	0	6.0	0	–	No fertilizations
12	0.63	65.2	0	7.9	0	–	No fertilizations

Identification code (ID) refers to ID of Table 9.1
*Not available

the Turin site (site 7 and 8). The overall mean EF was 0.7 % ± 2.3 (2 SD) and 0.3 % ± 0.4 (2 SD) including or excluding this site from the analysis, respectively. For a better characterization of the magnitude of N_2O emissions, the total fertilizer induced emission (FIE), the maximum peak value and the time length of the emission peak (back to baseline) were calculated (Table 9.2).

FIE for each fertilization event increased with increasing fertilizer N addition following an exponential relationship (Fig. 9.1) in accordance with Bouwamn et al. (2002b) observations. However, the rate of increase of N_2O flux per unit of

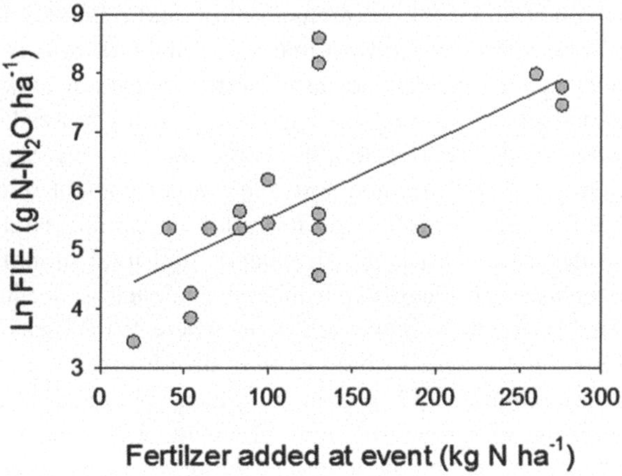

Fig. 9.1 Natural logarithm of fertilizer induced emission (FIE) of N_2O plotted versus amount of N fertilizer added at each single fertilization event. The fitting equation is $Y = 0.013\ X + 4.20$, $R^2 = 0.48$, $P = 0.001$

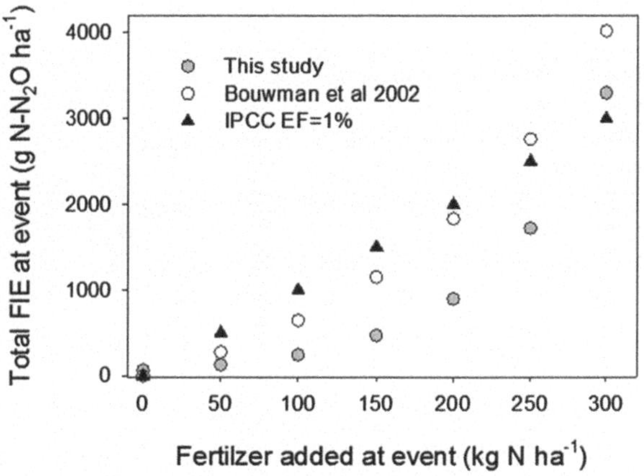

Fig. 9.2 Expected N_2O fluxes for increasing N fertilizer addition calculated using equation derived in Fig. 9.1, the IPCC EF 1 % and Bouwman et al. (2002b) equation

added fertilizer was much lower than the one predicted by the exponential relationship proposed by Bouwman et al. (2002b) (Fig. 9.2), assuming soil characteristics similar to those reported in Table 9.1. The maximum difference between the two approaches occurred at fertilizer concentrations between 150 and 250 kg N ha^{-1}. Hence, this means that the N_2O emissions predicted in this range of values using either IPCC EF or Bouwman et al. (2002b) approach would significantly overestimate (up to three folds) the real emissions in spring-summer maize crops when applied to Italian conditions. Bouwman et al. (2002a) data used to derive the empirical model and which IPCC EF is also based on, come from the analysis of 846 studies in humid tropics and temperate areas, where Mediterranean climate is almost never represented. Since the Italian sites have high rates of fertilization in line with other European agricultural crops, the average soil water content is most probably the main limiting factor for high N_2O emissions even during fertilization events. In the Mediterranean area the combination of high temperature, which stimulates evapotranspiration, and relatively low rainfall during the spring-summer periods, tends to maintain soils moisture well below 60 % of water filled pore space (WFPS), value of water saturation at which N_2O emissions increase exponentially in non limiting conditions of N availability (Davidson 1993). Even when crops are irrigated, the water input is calibrated for plant needs and, in particular for well drained soils, soil water content remains relatively low, except in the few minutes following irrigation or rainfall. Higher EFs might be expected for winter crops, when rain frequency and total monthly input increases while EPT decreases, although experimental evidence is needed to confirm this hypothesis. The Turin site was the sole field where WFPS was between 66 and 80 % during fertilization, which would explain the observed high fluxes.

Maximum N_2O emission values, obtained at flux peaks, were correlated with the magnitude of total FIE (Fig. 9.3), whereas an inverse relationship was observed between the length of the fertilizer induced emission period and the maximum flux peak (Fig. 9.4). Data indicate that high emissions are obtained only at

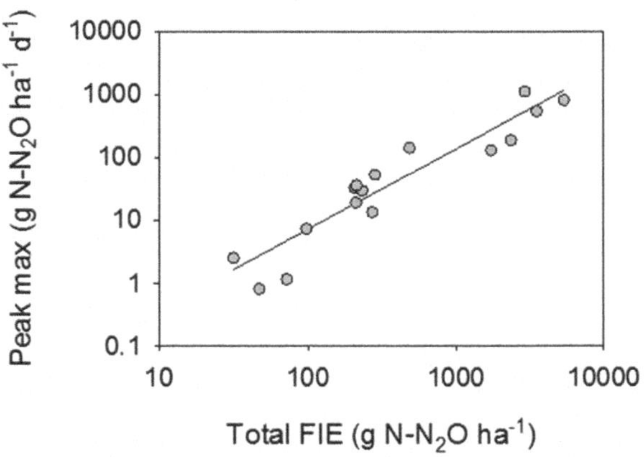

Fig. 9.3 Maximum emission rate (g N_2O–N ha^{-1} d^{-1}) versus total fertilizer induced emission (g N_2O–N ha^{-1})

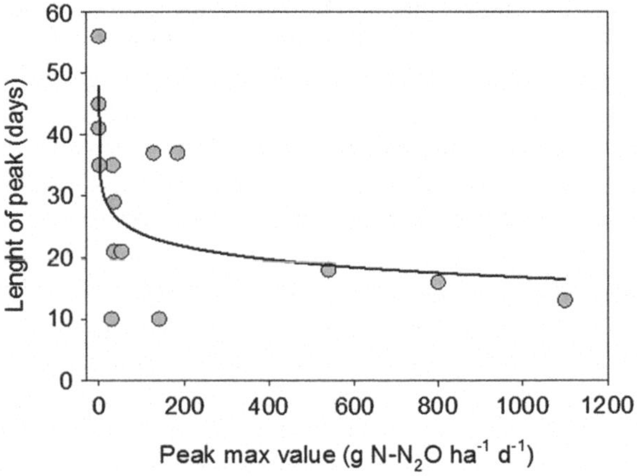

Fig. 9.4 Length of the fertilizer induced emission period (days) versus maximum N_2O emission rate (g N_2O–N ha^{-1} d^{-1})

high fertilization rates, with N_2O flush which concentrates on few days of very high emissions (<20). At intermediate values of fertilization the average emitting period lasts slightly more than 30 days arriving at a maximum of 56 days.

9.5 Conclusions

The difference in EFs found between data obtained in Italian experimental sites of maize and the data calculated using IPCC' and Bouwman et al. (2002b) empirical equation, suggests that climate might have a stronger influence on N_2O emissions rates than soil characteristics, for a given input of Nr. This poses a clear problem in terms of availability of representative data for specific climatic areas.

The few available data from Bouwman dataset, which are derived from Mediterranean climate sites, report an EF of 0.28 % for Spain (Andalusia 3 sites, Catalonia 3 sites) and 0.5 % for 3 sites in Modena. Both values are closer to the presented estimates on maize (0.3–0.7 %) rather than to the 1 % by IPCC average estimates. Averaging the above reported estimates for Mediterranean climate sites, we might indicate an average EF of 0.45 % $\pm$ 0.2 as a more realistic estimate than 1 % for most of the crops of the Italian peninsula excluding areas which might already belong to climatic areas not specifically classified as Mediterranean.

However, the present study has also further evidenced that the relationship between Nr addition and FIE is not linear, so that the EF will change for increasing Nr addition rates. The empirical equation calculated in this study (Fig. 9.1) could then be applied to calculate the FIE rather than a simple EF when climatic conditions resemble those encountered in this study. This would take into account both the climatic impact and the fertilizer addition rate.

References

Alluvione F, Bertora C, Zavattaro L, Grignani C (2010) Nitrous oxide and carbon dioxide emissions following green manure and compost fertilization in corn. Soil Sci Soc Am J 74:384–395

Bouwman AF, Boumans LJM, Batjes NH (2002a) Emissions of N_2O and NO from fertilized fields: summary of available measurement data. Global Biogeochem Cycles 16(4):1058–1071

Bouwman AF, Boumans LJM, Batjes NH (2002b) Modeling global annual N_2O and NO emissions from fertilized fields. Global Biogeochem Cycles 16(4):1080–1089

Castaldi S (2000) Responses of nitrous oxide, dinitrogen and carbon dioxide production and oxygen consumption to temperature in forest and agricultural light-textured soils determined by model experiment. Fertil Biol Soils 32:67–72

Castaldi S, Carfora A, Fiorentino A, Natale A, Messere A, Miglietta F, Cotrufo MF (2009) Inhibition of net nitrification activity in a Mediterranean woodland: possible role of chemicals produced by Arbutus unedo. Plant Soil 315:273–283

Castaldi S, Riondino M, Baronti S, Esposito FR, Marzaioli R, Rutigliano FA, Vaccari FP, Miglietta F (2011) Impact of biochar application to a Mediterranean wheat crop on soil microbial activity and greenhouse gas fluxes. Chemosphere 85(9):1464–1471

Denman K, Brasseur G, Chidthaisong A, Ciais P, Cox PM, Dickinson RE, Hauglustaine D, Heinze C, Holland E, Jacob D, Lohmann U, Ramachandran S, Da Silva Dias PL, Wofsy SC, Zhang X (2007) Couplings between changes in the climate system and biogeochemistry. In: Solomon S, Qin D, Manning M, Chen Z, Marquis M, Averyt KB, Tignor M and Miller HL (eds) Climate change 2007: the physical science basis. contribution of working group I to the fourth assessment report of the Intergovernmental Panel on Climate Change. Cambridge University Press, Cambridge

Davidson EA (1993) Soil water content and the ratio of nitrous to nitric oxide emitted from soil. In: Oremland RS (ed) Biogeochemistry of global change: radiatively active trace gases. Chapman & Hall, London, pp 369–386

Fierro A, Forte A (2012) Measurements of CO_2 and N_2O emissions in the agricultural field experiments of the MESCOSAGR project (Chap. 9). In: Piccolo A (ed) Carbon sequestration in agricultural soils. Springer, Berlin

Galloway JN, Townsend AR, Erisman JW, Bekunda M, Cai Z, Freney JR, Martinelli LA, Seitzinger SP, Sutton MA (2008) Transformation of the nitrogen cycle: recent trends, questions, and potential solutions. Science 320:889–892

Hutchinson G, Mosier AR (1981) Improved soil cover method for field measurements of nitrous oxide fluxes. Soil Sci Soc Am J 45:311–316

IPCC (1997) Houghton JT, Meira Filho LG, Lim B, Tréanton K, Mamaty I, Bonduki Y, Griggs DJ, Callander BA (eds) Revised 1996 IPCC Guidelines for National Greenhouse Inventories. IPCC/OECD/IEA, Paris, France

IPCC (2006) Guidelines for National Greenhouse Gas Inventories, Prepared by the National Greenhouse Gas Inventories Programme. In: Eggleston HS, Buendia L, Miwa K, Ngara T, Tanabe K (eds) IGES, Japan

IPCC (2007) Climate Change 2007: The Physical Science Basis. Contribution of Working Group I to the Fourth Assessment Report of the Intergovernmental Panel on Climate Change. In: Solomon S, Qin D, Manning M, Chen Z, Marquis M, Averyt KB, Tignor M, Miller HL (eds) Cambridge University Press, Cambridge, United Kingdom and New York, NY, USA, pp 996

Ranucci S, Bertolini T, Vitale L, Di Tommasi P, Ottaiano L, Oliva M, Amato U, Fierro A, Magliulo V (2011) The influence of management and environmental variables on soil N_2O emissions in a crop system in Southern Italy. Plant Soil 343:83–96

Ravishankara AR, Daniel JS, Portmann RW (2009) Nitrous oxide (N_2O): the dominant ozone-depleting substance emitted in the 21st century. Science 326:123–125

Rees RM, Augustin J, Alberti G, Ball BC, Boeckx P, Cantarel A, Castaldi S et al (2013) Nitrous oxide emissions from European agriculture; an analysis of variability and drivers of emissions from field experiments. Biogeosciences 10:2671–2682

Smith KA (1990) Anaerobic zones and denitrification in soil: modelling and measurement. In: Revsbch NP and Sørensen (eds) Denitrification in soil and sediments. Plenum Press, New York, 228–240

Smith KA, Clayton H, McTaggart IP, Thomson PE, Arah JRM, Scott A (1995) The measurement of nitrous oxide emissions from soil by using chambers. Philos Trans R Soc A 351:327–338

Chapter 10
Cropland and Grassland Management

Emanuele Lugato, Agata Novara, Damiano Gianelle, Loris Vescovo and Alessandro Peressotti

Abstract According to the latest National Inventory, the Italian agricultural sector is a source of GHGs with 34.5 Mt of CO_2 eq in 2009, corresponding to 7 % of the total emissions (excluding LULUCF). In particular, more than half (19.1 Mt of CO_2 eq) are N_2O emissions from soils. Although the national methodology is in accordance with Tier 1 and 2 approaches proposed by the IPCC (2006), still empirical emission factors are used to assess the emission from fertilizer (e.g. 0.0125 kg N_2O–N kg^{-1} N from synthetic fertilizers). Disaggregated data at sub-national level, including models and inventory measurement systems required by higher order methods (i.e. Tier 3), are not available in Italy so far and comparisons with the other two approaches cannot be performed at the moment. Despite the large soil organic carbon pool in the agricultural soils and the recent institutionalization of the 'National Registry for Carbon sinks' by a Ministerial Decree on 1st April 2008, the last Italian greenhouse gas Inventory did not report CO_2 emissions from the agricultural sector. In this context, this chapter wants to summarize the main outcomes coming from the

E. Lugato (✉)
Institute of Biometeorology (IBIMET-CNR), Via G.Caproni 8, 50145 Firenze, Italy
e-mail: e.lugato@ibimet.cnr.it

A. Novara
Agricultural and Forest Sciences Department, University of Palermo,
Viale delle Scienze, Palermo, Italy

D. Gianelle
Department of Agro-Environmental System, University of Palermo, Palermo, Italy

L. Vescovo
Department of Sustainable Agro-Ecosystems and Bioresources,
Research and Innovation Centre, Fondazione Edmund Mach, 38010 S. Michele All'adige,
Trento, Italy

A. Peressotti
Department of Agricultural and Environmental Sciences, University of Udine,
Via delle Scienze 206, 33100 Udine, Italy

R. Valentini and F. Miglietta (eds.), *The Greenhouse Gas Balance of Italy*,
Environmental Science and Engineering, DOI 10.1007/978-3-642-32424-6_10

main long-term experiments present in Italy by integrating experimental and modeling approaches, which can provide national emission rates and a solid base to test and calibrate simulation models to estimate greenhouse gases emissions from Italian agricultural soils. What emerges clearly from the analysis is that the agro-ecosystems may sequester large amount of SOC if appropriate management practices are adopted. Moreover, the use of simulation models calibrated at local level and spatially applied, as done for the Carboitaly project, may certainly reduce the uncertainty of these estimations.

10.1 Measuring and Modeling GHG Balance in Agriculture: From Field to Territory

10.1.1 Field Experiments and Chronosequences

SOC and nitrogen cycles in agricultural soils may be strongly variable both in time and space, therefore the study of their dynamics often requires different levels and approaches of investigation.

Measurements of CO_2 fluxes from agro-ecosystems are often too short to detect non-linear and long-term trends related to land use and management change. In fact, SOC has an average mean residence time in the order of years (Kuzyakov 2006), hence the effect of agricultural practices on C stock may be better analyzed in long-term trials. These experiments can provide specific country values that may help to improve the accuracy of the National Inventory Report (NIR) by the implementation of a tier 2–3 approach.

Fortunately, some long-term experiments have been set-up in Italy in the past, providing results that could be briefly reviewed. Morari et al. (2006) and Lugato et al. (2007) published data of a long-term trial began in 1962 at the experimental farm of the University of Padova, located in Legnaro (NE Italy, 5026115 m N, 732390 m E). Another old trial, started in 1966 and located in Cadriano (BO–4935649 m N, 686661 m E), was extensively analyzed by Triberti et al. (2008). Furthermore, Mazzoncini et al. (2011) reported interesting results of a more recent experiment, established in 1993 at the Interdepartmental Centre for Agro-Environmental Research (4835964 m N; 606152 m E) by the University of Pisa. All these trials have been carried out in field plots and are still undergoing.

The main outcomes of these experiments (Table 10.1) clearly evidenced as recommended management practices could increase SOC of agricultural fields, making the agriculture as an eligible sink under the future EU policies. In particular, the change of land use from cropland to grassland, the no-till adoption and the use of farm yard manure showed the higher sequestration rates, ranging from 0.26 to 0.67 $t\ C\ ha^{-1}\ y^{-1}$. The crop rotation had a marginal effect compared to monoculture in Legnaro, while similar indications came from residues management, both from trials conducted in Legnaro and Pisa.

Table 10.1 Rate of C sequestration of recommended management practices (RMP) derived from the main long-term trials in Italy compared to the business as usual scenario (BAU)

		Sequestration rate (t C ha^{-1})		
BAU	RMP	Legnaro (PD)*	Cadriano (BO)**	Pisa (PI)***
Crop rotations	Monoculture	0.02 (±0.001)		
Grass	Crop rotation	0.4 (±0.05)		
Residue incorporation	Residue removal	0.10 (±0.07)	0.16	
		0.17		
High inorganic rate	Low inorganic rate	0.038 (±0.016)		0.31–0.35[d]
		0.20[a] (±0.07)		
Animal Manure	Inorganic fertilizer	0.58[b] (±0.15)	0.26[b]	
		0.27[c] (±0.13)	0.18[c]	
Conventional tillage	No tillage			0.67[e]
Main crops	Cover crops			0.08–0.34[e]

*Morari et al. (2006) from Table 6 and Lugato et al. (2006); **Triberti et al. (2008) from Table 4; ***Mazzoncini et al. (2011) from Table 7

[a]In wheat monoculture; [b]farm yard manure; [c]slurry; [d]calculated as N1 and N2–N0 rate; [e]calculated as RMP—control rate

Beside these long-term trials, other shorter but very well monitored experimental fields are present in Italy, such as that carried on by the University of Udine at Beano (see Sect. 2.2.2). This site is equipped with two Eddy Covariance stations and a soils respiration system with several dynamic chambers, which also allow the measurement of N_2O fluxes (Alberti et al. 2010). The fluxes campaign showed that conversion from maize to alfalfa was not effective in sequestering SOC with respect to continuous maize, in the short-term (2 years). However, data coming from a land use change chronosequence in that area (NE—Italy) (del Galdo et al. 2003), clearly highlighted that permanent grassland contained almost double SOC content than maize in the first 10 cm, and the same amount of C in the deeper layer (10–30 cm). In this study, afforestation of cultivated soils occurred 20 years ago resulted in significant sequestration of new C and stabilization of old C in physically protected soil organic matter fractions.

10.1.2 Modeling Application and Territorial Upscaling

Long and short-term field experiments allow the investigation of C and N turnover on a different temporal scale, but the results upscaling may be limited by the representativeness of the site pedo-climatic condition. In this context, biogeochemical process-based models could become very powerful tools, both to make prediction in time and space.

Up to present, with reference to modeling application in Italian agricultural fields only few works have been published. Lugato et al. (2006, 2007) extensively tested the SOC Century model using the long-term dataset of the University of Padova. The model was very accurate in simulating different agricultural managements involving a wide range of crop rotations and type of organic and mineral fertilizations. Furthermore, the authors made an analysis of the scenario (Lugato and Berti 2008), showing that recommended management practices could affect the C balance more strongly than climate change. Conversion to grassland was the most promising practice for sequestering C, allowing an average accumulation of 2.5 and 14 t C ha^{-1} at the end of the first Kyoto commitment period (2012) and in 2080, respectively.

The long-term dataset of the University of Bologna (Cadriano) was used by Plaza et al. (2012) to evaluate the performances of the process-based C model CQUESTR. The results indicated that the model was able to predict the SOC trend with a reasonable degree of accuracy, but further improvements may result from a better parameterization of the chemical composition and degradability of organic amendments.

Despite other model applications have been applied locally also in Mediterranean areas (as illustrates specifically in the next paragraph), territorial simulations at high spatial resolution are still lacking in Italy. As a deliverable of the project Carboitaly, Lugato et al. (2010) developed a platform of simulation to run DNDC for the entire national territory on a grid of 1 km^2, linking the model with geographical databases. The main outcomes showed that N_2O emissions were generally low (<0.5 kg N ha^{-1}) with a simulated emission factor from fertilizers of 0.008, lower than the IPCC default of 0.0125 kg N_2O–N kg^{-1} N. Cumulative emissions, covering 52 % of the arable land and 26 % of the woodland crops simulated, were 1.52 (±0.04) and −0.08 (±0.001) Mt of CO_2 eq for N_2O and CH_4, respectively (negative values indicate CH_4 oxidation).

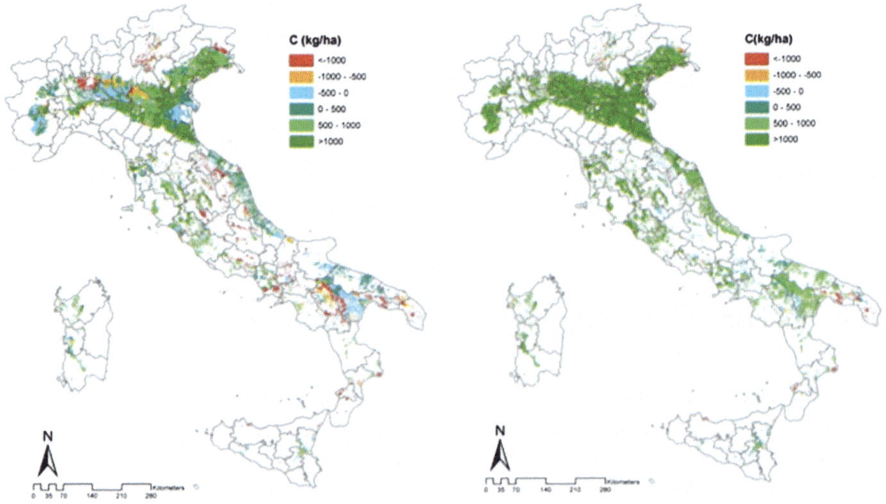

Fig. 10.1 Annual CO_2 fluxes (kg C ha^{-1}) in the grain maize with alternative management practices: conventional (*left*) versus reduced tillage (*right*)

This platform allowed the possibility to assess the net effects of alternative management practices applied spatially to different crops as, for example, the adoption of a minimum tillage on grain maize (Fig. 10.1). Under this scenario, the minimum tillage adoption allowed the accumulation of about 1.1 t ha^{-1} of SOC on average, saving more than 4 Mt of CO_2 eq considering also the other GHG.

Another attempt to investigate the effect of the change in land use on SOC stock at very high resolution was made by Coslovich (2011). He applied the GEF-SOC system (Easter et al. 2007), which links Century model with GIS data, to the administrative unit of Padova (NE Italy). Reconstructing the past land use since 1800, he was able to track the SOC evolution during the last century, making also a prediction of the future values according to two climatic scenarios (Had3A1FI and PCM-B1) (Fig. 10.2).

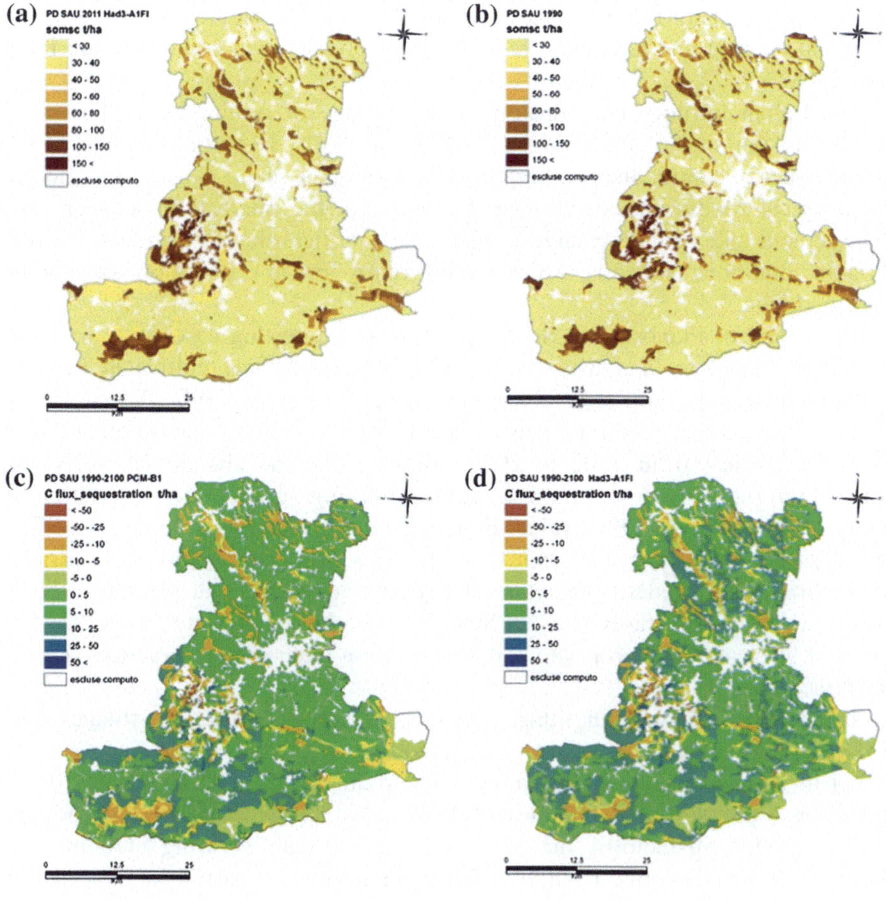

Fig. 10.2 SOC stock (t C ha^{-1}) simulated for the year 2011 (**a**) and 1990 (**b**) in province of Padova, and SOC variation (2100–1990) for two climatic scenarios PCM-B1 (**c**) and Had3A1FI (**d**)

10.2 SOC Changes in Relation to Cropping Systems and Management in Mediterranean Semi-arid Areas

Understanding SOC changes in relation to ecosystem management is needed to develop C accounting models and future policies, oriented at attenuating the GHG warming effect. In arable lands, optimum levels of SOC could be mainly managed through crop rotations, tillage methods, inorganic fertilizers and organic manures (Kundu et al. 2007; Lal 2004). To mitigate the universally admitted deleterious effect of conventional cropping on SOC, several studies suggested to increase the frequency in the rotation of high biomass production and crops with elevated C/N ratio residue, as well as fertilize, apply organic manure, irrigate, and/or reduce tillage intensity (Su et al. 2006; Studdert and Echeverria 2000; Lal 2004; Sainju et al. 2006; Morari et al. 2006). However, the magnitude in SOC change following such usually recommended practices is soil and site specific. The potential of different eco-regions of the world to sequester C is in fact climatic dependent, being higher in tropical and temperate regions where crop growth conditions are more favourable.

The Italian territory encompasses a wide range of orographic and climatic conditions, ranging in latitude from about 47° to 36°N. In particular, most of the coastal areas of central and southern Italy, including the major islands, are affected by a typical Mediterranean climate. In this area, the interaction between land management and climate change is particularly critical due to frequently severe drought conditions and soils, which are inherently low in C and more susceptible to degradation.

The Island of Pianosa (see Sect. 2.2.18) is an interesting case study for SOC evolution related to land use change, as it was intensively cultivated for several centuries and completely abandoned since the end of 1990s. Vaccari et al. (2012) conducted a medium-term campaign of CO_2 fluxes measurement with Eddy Covariance from 2002 to 2009, showing that the abandoned ecosystem was able to rapidly recover the SOC depleted during agricultural exploitation by about 30 % above the 1990s value. Furthermore, the authors calibrated the SOC-ecosystem model Century (Parton et al. 1988) with the measured data, in order to perform some scenario analysis. The results suggested that re-cultivation of these soils may potentially cause a SOC loss ranging from 41 to 58 t C ha^{-1} in the next 90 years, if appropriate soil conservation management practices are not implemented.

The interaction between tillage practices (conventional vs. no-tillage) with different rates of N fertilizer was measured and modelled by (de Sanctis et al. 2012) in another typical Mediterranean environment: the experimental farm of University of Sassari in Agugliano (4821339 m N, 368026 m E). In this study the DSSAT model satisfactorily matched the measured data, showing SOC increase with no till and moderate to high N fertilization rates compared to conventional tillage. The application of these techniques suggested a steadily increase of SOC over a 50-year period at a rate of about 0.3 t ha^{-1} y^{-1}.

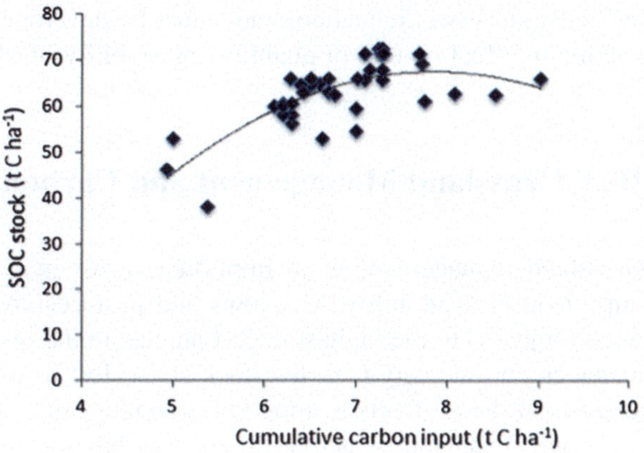

Fig. 10.3 Relationship between cumulative C input and stock, in two long-term experiments in semiarid environment

Indeed, contrasting observations came from two long-term experiments in Sicily (Pietranera farm: 4155073 m N, 368943 m E; and Sparacia farm: 4164079 m N, 385269 m E), where no linear correlation between cumulative carbon input and soil carbon accumulation was found (Fig. 10.3). The annual SOC sequestration increased with a cumulative C input up to 7.4 t C ha^{-1}, then showing a saturation threshold with the actual cultivation practices and pedo-climatic conditions at the investigated sites (Barbera et al. 2011).

In semiarid cropping systems, different steady state levels can be reached by crop rotation management due to different mineralization rate of plant residue, as emerged from these long-term trials. Surprisingly, the higher SOC content in bulk soil was recorded in cereal monoculture in comparison to cereal-leguminous rotations, while there were no significant differences between crop rotations in the smallest soil aggregate fraction (<25 μm) (Barbera et al. 2011). These results are in disagreement with other studies that reported the positive effects on enhancing SOC in crop rotations including leguminous species (Upendra et al. 2005; Ashraf et al. 2004). The lowest efficiency of C retained from crop residue with a low C/N ratio and less rich in lignin (crop rotation with leguminous species) was likely due to the higher labile C fraction than the sole wheat residues.

In all crop rotations analysed, the effect of soil management was no significant different. Although many authors claim that conservation tillage (reduced and no-till practices) increase SOC in the surface layer (Sainju et al. 2006; Melero et al. 2009; López-Bellido et al. 2010), improves soil aggregation and preserves soil resources better than conventional tillage practices (Six et al. 2004), SOC values did not significantly differ between conventional tillage (CT) and no-tillage (NT), during observation carried out in these long-term experiments.

Despite the RMPs suggested by IPCC, the efficiency of carbon sequestration in semiarid environment was apparently low, due to hot and drought climate which facilitates the rapid oxidation of organic C. Conventional cropping management did not result in SOC depletion, since the most of C was stored and stabilized in the more stable silt-clay fraction. As highlighted from these long-term experiments in Sicily, site-specific technologies must be developed and validated to obtain the maximum effect in term of quantity and stability of SOC sequestered.

10.3 Grassland Management and Carbon Balance

Grassland management is an important factor in C balance also because of its impact on N availability. Meadows and pastures are among the ecosystems that were subjected to the highest rate of change in the last decades for the reduction of management intensity (e.g. livestock stock, fertilization, abandonment), hence the analysis of these effects is important to understand and predict SOC stock dynamics. Land management activities, such as haying, fertilisation and grazing affect the assimilation and ecosystem respiration through changes in photosynthetic activity and quality of respiratory substrate. The removal of aboveground biomass modifies the soil microclimate (temperature and humidity), then influencing the soil respiration. Moreover, mineral and organic fertilizations affect the nutrient availability and thus leaf photosynthesis and ecosystem respiration components. Among agronomic management techniques, conversion from cropland and irrigation cause the largest increase in C sequestration rate (Conant et al. 2001); furthermore, well-managed pastures can have a positive impact on SOC level, accumulating up to 7 % with respect to no grazed areas. Timing and frequency of management practices in grasslands could affect the more labile C pools, leading to short-term ecosystem response that can be detected by fast measurements, such as Eddy Covariance (Wohlfahrt et al. 2008).

In the experimental area of Monte Bondone (see Sect. 2.2.13) an analysis of the impact of climate on C balance of an alpine grassland was carried out using an Eddy Covariance station. It is worth noting that the observation years (2003–2009) were characterised by a high variability in precipitation, covering about 80 and 55 % of the historical range observed on a yearly basis and during the growing season, respectively. The results showed that the portion of inter-annual variability (IAV) generated by climate variability is considerably smaller than that induced by changes of ecosystem responses, and the relative importance of climate related IAV decreases from gross primary production-GPP (36 %) to total ecosystem respiration-TER (22 %) and to net ecosystem exchange-NEE (20 %), suggesting a strong resistance of ecosystem carbon fluxes to the direct effect of climate drivers. This analysis only marginally supports the hypothesis that temperature limited ecosystems are highly sensitive and vulnerable to climate change though it rather stresses the effectiveness of acclimation in limiting the impact of climate variability on ecosystem productivity (Marcolla et al. 2011).

Considering that the inter-annual variability of the climate does not seem to have a strong impact on the carbon balance of grasslands studied, an experimental campaign was carried out on the Monte Bondone plateau (2 km^2) in which different systems of grassland management are present (Fig. 10.4). Meadows represent the principal land use on this plateau and they are traditionally managed with low

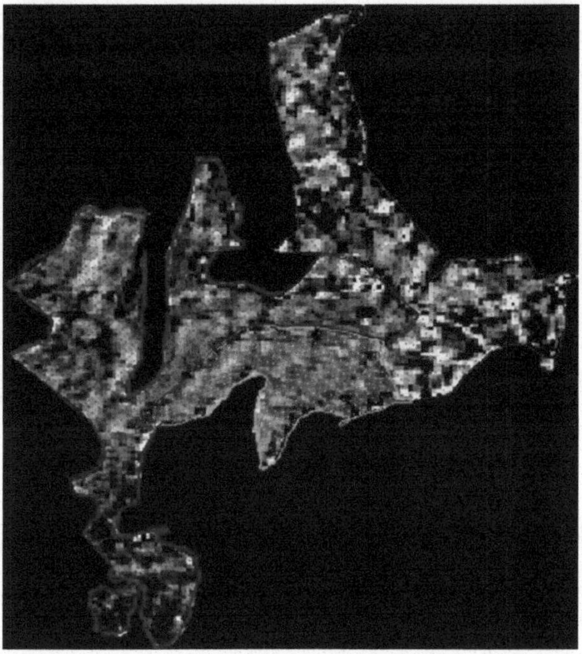

Fig. 10.4 Map of *above ground* biomass in grassland area of Viote Monte Bondone (pasture, *blue area*; intensive managed meadow *green area*; extensive managed meadow *red area*)

mineral fertilization and cut once a year in mid-July. Some areas are not fertilized and biomass is commonly harvested as hay. Pastures are located on the southern parts of this area and grazed between June and September by cattle. There are some areas that have been abandoned and have been invaded by shrubs. A peat bog is located in the central part of the plateau.

Remote and proximal sensing tools were used to upscale plot measurements to ecosystem level. Grassland biophysical parameters, linked to carbon stock, such as LAI, biomass, phytomass, and Green herbage ratio (GR) were measured using both ground-truth and satellite data. The performances of the most common Near Infrared Radiometer (NIR) and red/green-based vegetation indices were compared with ones that also make use of the Microwave Imaging Radiometer (MIR) band, in relation to their ability to predict grassland biophysical parameters. From satellite platforms, an IRS-1C-LISS III image (25 m resolution in the visible-NIR and 70 m resolution in the MIR) and a SPOT 5 image (10 m resolution in the visible-NIR and MIR) were used. MIR-based indices performed better than NIR and red/green-based ones in estimating biophysical variables, with no saturation effect (Vescovo and Gianelle 2008). Biomass showed a linear regression with Canopy Index (MIR/green ratio) and with the Normalised Canopy Index (NCI) calculated as a normalised difference between MIR and green bands (IRS: R2 = 0.91 and 0.90, respectively; SPOT: R2 = 0.63 and 0.64). The use of satellite images, in particular IRS ones, permitted to reconstruct the grassland biomass of the Monte Bondone plateau, and to analyze the effect of the different management practices and intensity. The pasture and the extensive managed grassland produced the same biomass (310.3 ± 104.9 vs. 303.9 ± 130.5 g m^{-2}) while higher values were found in the intensive managed grassland (356.5 ± 179.6).

10.4 Conclusions

Large quantities of information are coming from the network of Eddy Covariance stations and long-term agronomical experiments located in Italy. The time is rape to integrate all that information into modeling tools, able to improve the national GHG balance in agriculture. Given the clear orientation of EU policies in achieving a more sustainable and green economy, the agricultural sector will also be required to take part into this process. The use of simulation models calibrated at local level and spatially applied, as per the Carboitaly project, may certainly help in designing the best 'agricultural systems', towards a more sustainable agriculture.

Acknowledgments We would like to thank prof. L. Gristina (University of Palermo) for his valuable contribution to Sect. 10.2.

References

Alberti G, Vedove GD, Zuliani M et al (2010) Changes in CO_2 emissions after crop conversion from continuous maize to alfalfa. Agric Ecosyst Environ 136:139–147. doi:10.1016/j.agee.2009.12.012

Ashraf M, Mahmood T, Azam F, Qureshi RM (2004) Comparative effects of applying leguminous and non-leguminous green manures and inorganic N on biomass yield and nitrogen uptake in flooded rice (Oryza sativa L.). Biol Fert Soils 40:147–152

Barbera V, Poma I, Gristina L, Novara A, Egli M (2011) Long-term cropping systems and tillage management effects on soil organic carbon stock and steady state level of C sequestration rates in a semiarid environment. Land Degrad Dev 23:82–91

Conant RT, Paustian K, Elliott ET (2001) Grassland management and conversion into grassland: effects on soil carbon. Ecol Appl 11:343–355

Coslovich D (2011) Stima del sequestro potenziale di carbonio nei suoli agricoli del Veneto con il GEFSOC Modelling System. PhD thesis

de Sanctis G, Roggero PP, Seddaiu G, Orsini R, Porter CH, Jones JW (2012) Long-term no tillage increased soil organic carbon content of rain-fed cereal systems in a Mediterranean area. Eur J Agron 40:18–27

del Galdo I, Six J, Peressotti A, Cortufo F (2003) Assessing the impact of land-use change on soil C sequestration in agricultural soils by means of organic matter fractionation and stable C isotopes. Glob Change Biol 9:1204–1213. doi:10.1046/j.1365-2486.2003.00657.x

Easter M, Paustian K, Killian K et al (2007) The GEFSOC soil carbon modelling system: a tool for conducting regional-scale soil carbon inventories and assessing the impacts of land use change on soil carbon. Agric Ecosyst Environ 122:13–25. doi:10.1016/j.agee.2007.01.004

IPCC (2006) Guidelines for National Greenhouse Gas Inventories, Prepared by the National Greenhouse Gas Inventories Programme. In: Eggleston HS, Buendia L, Miwa K, Ngara T, Tanabe K (eds). IGES, Japan

Kundu S, Bhattacharyya R, Prakash V, Ghosh BN, Gupta HS (2007) Carbon sequestration and relationship between carbon addition and storage under rainfed soybean–wheat rotation in a sandy loam soil of the Indian Himalayas. Soil Till Res 92:87–95

Kuzyakov Y (2006) Sources of CO2 efflux from soil and review of partitioning methods. Soil Biol Biochem 38:425–448. doi:10.1016/j.soilbio.2005.08.020

Lal R (2004) Soil carbon sequestration to mitigate climate change. Geoderma 123:1–22

López-Bellido RJ, Montan JM, López-Bellido FJ, López-Bellido L (2010) Carbon sequestration by tillage, rotation, and nitrogen fertilization in a Mediterranean Vertisol. Agron J 102:310–318

Lugato E, Berti A (2008) Potential carbon sequestration in a cultivated soil under different climate change scenarios: a modelling approach for evaluating promising management practices in north-east Italy. Agric Ecosyst Environ 128:97–103. doi:10.1016/j.agee.2008.05.005

Lugato E, Berti A, Giardini L (2006) Soil organic carbon (SOC) dynamics with and without residue incorporation in relation to different nitrogen fertilisation rates. Geoderma 135:315–321. doi:10.1016/j.geoderma.2006.01.012

Lugato E, Paustian K, Giardini L (2007) Modeling soil organic carbon dynamics in two long-term experiments of north-eastern Italy. Agric Ecosyst Environ 120:423–432. doi:10.1016/j.agee.2006.11.006

Lugato E, Zuliani M, Alberti G et al (2010) Application of DNDC biogeochemistry model to estimate greenhouse gas emissions from Italian agricultural areas at high spatial resolution. Agric Ecosyst Environ 139:546–556. doi:10.1016/j.agee.2010.09.015

Marcolla B, Cescatti A, Manca G, Zorer R, Cavagna M, Fiora A, Gianelle D, Rodeghiero M, Sottocornola M, Zampedri R (2011) Climatic controls and ecosystem responses drive the inter-annual variability of the net ecosystem exchange of an alpine meadow. Agric For Meteorol 151:1223–1243

Mazzoncini M, Sapkota TB, Barberi P, Antichi D, Risaliti R (2011) Long-term effect of tillage, nitrogen fertilization and cover crops on soil organic carbon and total nitrogen content. Soil Till Res 114:165–174. doi:10.1016/j.still.2011.05.001

Melero S, López-Garrido R, Murillo JM, Moreno F (2009) Conservation tillage: Short- and long-term effects on soil carbon fractions and enzymatic activities under Mediterranean conditions. Soil Till Res 104:292–298

Morari F, Lugato E, Berti A, Giardini L (2006) Long-term effects of recommended management practices on soil carbon changes and sequestration in north-eastern Italy. Soil Use Manag 22:71–81. doi:10.1111/j.1475-2743.2005.00006.x

Parton WJ, Stewart JWB, Cole CV (1988) Dynamics of C, N, P and S in grassland soils: a model. Biogeochemistry 5:109–131

Plaza C, Gollany HT, Baldoni G et al (2012) Predicting long-term organic carbon dynamics in organically amended soils using the CQESTR model. J Soils Sedim 12:486–493. doi:10.1007/s11368-012-0477-1

Sainju UM, Singh BP, Whitehead WF, Wang S (2006) Carbon supply and storage in tilled and nontilled soils as influenced by cover crops and nitrogen fertilization. J Environ Qual 35:1507–1517

Six J, Bossuyt H, Degryze S, Denef K (2004) A history of research on the link between (micro) aggregates, soil biota, and soil organic matter dynamics. Soil Till Res 79:7–31. doi:10.1016/j.still.2004.03.008

Studdert GA, Echeverria HE (2000) Crop rotations and nitrogen fertilization to manage soil organic carbon dynamics. Soil Sci Soc Am J 64:1496–1503

Su YZ, Wang F, Suo DR, Zhang ZH, Du MW (2006) Long-term effect of fertilizer and manure application on soil-carbon sequestration and soil fertility under the wheat-wheat-maize cropping system in northwest China. Nutr Cycl Agroecosys 75:285–295. doi:10.1007/s10705-006-9034-x

Triberti L, Nastri A, Giordani G et al (2008) Can mineral and organic fertilization help sequestrate carbon dioxide in cropland? Eur J Agron 29:13–20. doi:10.1016/j.eja.2008.01.009

Upendra MS, Wayne FW, Bharat PS (2005) Biculture legume–cereal cover crops for enhanced biomass yield and carbon and nitrogen. Agron J 97:1403–1412

Vaccari FP, Lugato E, Gioli B et al (2012) Land use change and soil organic carbon dynamics in Mediterranean agro-ecosystems: the case study of Pianosa Island. Geoderma 175–176:29–36. doi:10.1016/j.geoderma.2012.01.021

Vescovo L, Gianelle D (2008) Using the MIR bands in vegetation indices for the estimation of grasslands biophysical parameters from satellite remote sensing in the Alps region of Trentino (Italy). Adv Space Res 41:1764–1772

Wohlfahrt G, Anderson-Dunn M, Bahn M, Balzarolo M, Berninger F, Campbell C, Carrara A, Cescatti A et al (2008) Biotic, abiotic and management controls on the net ecosystem CO_2 exchange of European mountain grassland ecosystems. Ecosystems 11:1338–1351

Part IV
Regional Case Studies

Chapter 11
The Role of Vineyards in the Carbon Balance Throughout Italy

Damiano Gianelle, Luciano Gristina, Andrea Pitacco, Donatella Spano, Tommaso La Mantia, Serena Marras, Franco Meggio, Agata Novara, Costantino Sirca and Matteo Sottocornola

Abstract A common belief is that agricultural fields cannot be net carbon sinks, but perennial tree crops, growing a permanent woody structure with a life cycle of decades could act as carbon sink. Vineyards are good candidates to test this hypothesis, because they are often grown with limited soil cultivation and produce plenty of woody pruning material that can be left on the ground. Three Eddy Covariance sites were established in different vineyards, along a north-south transect, in Italy, to study the role of vine cultivation in the carbon balance of the Italian penisula. The year 2009 was chosen as a reference year for the three sites, in order to compare carbon budget estimates in areas characterized by different meteorological, pedological and geomorphological conditions. In the three sites a

D. Gianelle (✉) · M. Sottocornola
Department of Sustainable Agro-Ecosystems and Bioresources, Research and Innovation Centre, Fondazione Edmund Mach, 38010 S. Michele all'Adige, Trento, Italy
e-mail: damiano.gianelle@fmach.it

L. Gristina · T. La Mantia · A. Novara
Agricultural and Forest Sciences Department, University of Palermo, Viale delle Scienze, Palermo, Italy

A. Pitacco · F. Meggio
Department of Agronomy, Food, Natural Resources, Animals and Environment, University of Padova, Viale dell'Università 16, 35020 Legnaro, PD, Italy

D. Spano · S. Marras · C. Sirca
Department of Science for Nature and Environmental Resources (DIPNET), University of Sassari, via E. de Nicola 9, 07100 Sassari, Italy

D. Spano · S. Marras · C. Sirca
Impacts on Agriculture, Forest and Natural Ecosystem Division (IAFENT), Euro-Mediterranean Center on Climate Changes (CMCC), Sassari, Italy

M. Sottocornola
Department of Chemical and Life Sciences, Waterford Institute of Technology, Waterford, Ireland

D. Gianelle
Foxlab Joint CNR-FEM Initiative, Via E. Mach 1, 38010 S. Michele all'Adige, Trento, TN, Italy

R. Valentini and F. Miglietta (eds.), *The Greenhouse Gas Balance of Italy*, Environmental Science and Engineering, DOI 10.1007/978-3-642-32424-6_11

carbon sink ranging between 814 (Negrisia site) and 89 (Serdiana site) g C m^{-2} y^{-1} was measured. Both climate (water availability and PAR) and management (in particular the presence of permanent grass cover) have a strong impact on the carbon balance of the ecosystems. Even if it can be argued that this sink may be only temporary and the built-up can be substantially disrupted at the end of the vineyard life cycle, these results show that there is a concrete possibility of storing carbon in agricultural soils. Proper practices can be defined to preserve this storage at best, greatly contributing to the global carbon budget.

11.1 Introduction

Croplands have been largely overlooked in terms of continuous carbon flux monitoring and carbon budget assessment. This is probably due to the common belief that agricultural fields cannot be net carbon sinks. Indeed, many technical inputs, heavy periodical harvests, and the repeated disturbances of the upper soil layers, all contribute to a substantial loss both of the old and newly-synthesized organic matter. Perennial tree crops, however, could behave differently: they grow a permanent woody structure, remain relatively undisturbed in the same field for decades, create a woody pruning debris, and are often grass-covered. Vineyards are good candidates to verify this hypothesis, because (at least in temperate climates) they are often grown with reduced or no soil cultivation, and are typically long-life crops that produce plenty of woody pruning material that is often left on the ground.

The global vineyard surface area shrunk in 2011 by 7.6 million hectares, with a loss of about 262 million hectares since 2000. The same trend with an even higher vineyard area loss was observed in Italy, with a decrease from 908,000 ha (2000) to 776,000 ha (2011, OIV 2012). Despite this, Italy today has the third worldwide vineyard acreage in the world and, alongside France, is the biggest world wine producer and the biggest exporter (around 40 % of the production). The grapevines are cultivated in every region of Italy, despite the large climatic differences along the peninsula, due to the plasticity and to the different varieties used in the cultivation. To understand the role played by vineyards in the national carbon budget, Eddy Covariance measurements in three vineyard sites along a north-south transect have been analyzed (Fig. 11.1). 2009 was chosen as a reference year for the three sites, in order to compare carbon budget estimates in areas characterized by different meteorological, pedological and geomorphological conditions. A crucial aspect in estimating the vineyard carbon balance at a national scale is to evaluate the impact of changes in land use management practices, and vineyard abandonment on the soil carbon stock. Changes in land use and management practices can easily affect the processes, which control the accumulation and stabilization of soil organic matter (SOM) and turn the soil from a C sink into a source of CO_2, while most of its functions are lost. On the global scale, inaccurate managements have a large impact on the atmospheric CO_2 concentration (IPCC 2001; Allmaras et al. 2000). In particular, it is well established that land-use change (LUC) is considered the second greatest

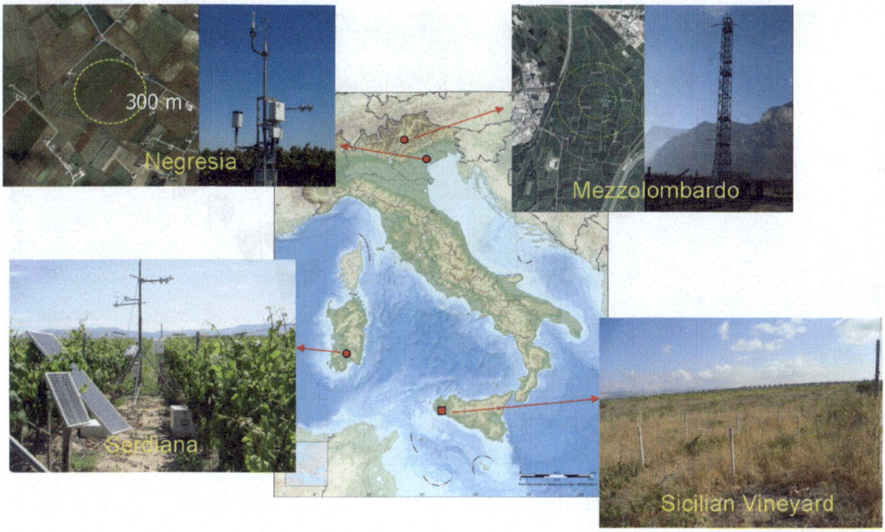

Fig. 11.1 Selected vineyard sites (*circle* Eddy Covariance sites; *square* Sicilian site for land use change studies)

cause of C emissions after fuel consumption (Watson et al. 2000). Over millennia, the vagaries and complexities of the social, political and economic conditions have transformed the Mediterranean basin ecosystems through land use and land use changes (Wainwright 1994; Grove 1996; Margaris et al. 1996). Moreover, industrialization and pressure from tourism activities during the last century led to an important socio-economic change in rural areas, due to land abandonment (Puigdefábregas and Mendizabal 1998; Thornes 1995). The area of permanent crops (vineyards, orchards) in Europe has therefore reduced substantially since the mid 1970s (Rounsevell et al. 2002). This was mostly due to the reduction between 1980 and 1995 of the vineyard covered area, which was strongly favoured by the European Community aid for the grubbing-up of vines. In order to analyze the effect of land use change on vineyard soil carbon stock, an area has been investigated in the south-west of Sicily where old vineyards were abandoned or replanted in the last decades (Fig. 11.1) (see Sect. 11.2).

11.1.1 Negrisia (Ponte di Piave, TV) Site

The FLUXNET station of Negrisia has been established in July 2005 in an extensive vineyard located close to Ponte di Piave (see Sect. 2.2.14 for a description). The vineyard is well suited for MODIS satellite platform observations. The station was intended for long-term monitoring of energy, water and CO_2 fluxes on a representative vineyard for North-Eastern Italy and has been operated continuously until 2011, with few gaps due to power failures or sensor maintenance. Leaf area

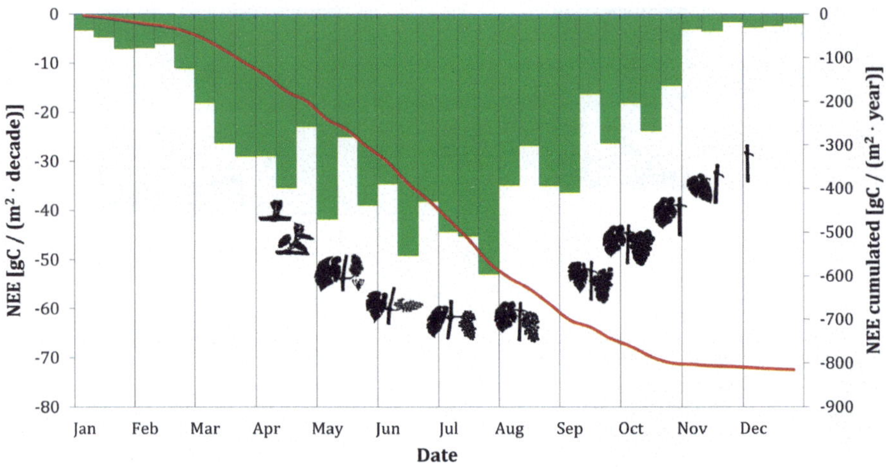

Fig. 11.2 Yearly course of decadal and cumulated NEE at the Negresia site, related to phenological stages

development was monitored regularly both by direct, destructive measurements on a sample of shoots, and by indirect, optical method, using a LiCor LAI-2000 canopy analyzer. Extensive biometrical sampling of leaves, shoots, trunks and roots has been carried out to provide independent measurements of vine growth dynamics and biomass partitioning.

The Eddy Covariance equipment is based on a 3-dimensional ultrasonic anemometer (Metek USA-1) and an open-path IR gas analyzer (LiCor LI-7500), both logged at 20 Hz by a low-power industrial PC. Ancillary measurements of net radiation (CNR-1, Kipp and Zonen) and basic meteorological parameters (air temperature and humidity, barometric pressure, rainfall, etc.) are measured every second and averaged every 30 min by a CR23X Campbell datalogger. Eddy Covariance data were processed according to the standard EUROFLUX methodology (Aubinet et al. 2000). Data quality check has been performed according to Foken and Wichura (1996) and gap-filling following Desai et al. (2005).

Following the set up of the station, the Negrisia vineyard proved to be a strong carbon sink. Data for 2009 (Fig. 11.2) showed a rapid increase of NEE already before budbreak, due to the increased activity of the grass cover, as soon as the temperature turned milder (Fig. 11.3). The quick development of the vine canopy further increased carbon absorption up to a maximum decadal NEE of 50 g C m^{-2}, which occurred around veraison. Leaf senescence and fall brought a steep decrease of NEE by the end of October. However, the system never showed a net release of carbon, because of the residual activity of the grass cover even during wintertime. Obviously, this behaviour strongly reflects both the adequate water availability (the vineyard never experienced significant stress, because of the generally high water table of the area) and the mild temperature (seldom below 0 °C). The annual total NEE (814 g C m^{-2}) suitably matched the biometrical measurements. Yield represented roughly 20 % of NEE and therefore Negrisia vineyard

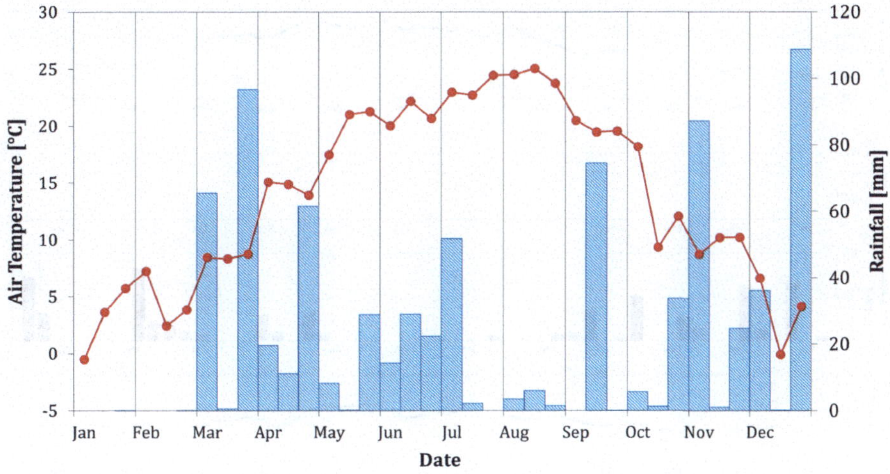

Fig. 11.3 Basic meteorological measurements for the site of Negrisia

behaved as a carbon sink even after harvest. Orchards and vineyards seem to be good candidates for mitigation practices, at least where an adequate water supply allows the establishment of a grass cover and cultivation is reduced or suppressed.

11.1.2 Valle dell'Adige (Mezzolombardo, TN) Site

The Mezzolombardo site is located in a vineyard in the Valle dell'Adige (Fig. 11.1; see Sect. 2.2.25 for a description).

Mass and energy fluxes have been continuously measured since 2008 at the site with an Eddy Covariance system mounted at a height of 7 m. The system consists of a Gill R3 3-D ultrasonic anemometer (Gill Instruments Ltd., Lymington, UK) and a Li-Cor 7500 open-path fast response infrared gas analyser (IRGA) (Li-Cor Inc., Lincoln, Nebraska, USA). Raw data are recorded at a frequency of 20 Hz and stored every 30 min into separate files with the Edisol software. Along with flux measurements, the tower is equipped with sensors for the measurement of standard meteorological parameters: short and long-wave radiation components (Kipp and Zonen CNR1, Delft, The Netherlands), photosynthetic active radiation (LiCOR 190 SA, Lincoln, Nebraska, USA), global solar radiation (LiCOR 200 SA, Lincoln, Nebraska, USA), direct and diffused radiation (BF3, Delta-T Devices, UK), air humidity, air temperature (Rotronic MP103A, Crawley, UK) and precipitation (Young 52202H, Traverse City, Michigan, USA). Soil temperature is measured at depths of 2, 5, 10, 20 and 50 cm (STP01, Hukseflux, Delft, The Netherlands), while volumetric soil water content is measured at 10 and 20 cm depths with CS615 reflectometers (Campbell Scientific Inc., Logan, Utah, USA). All meteorological variables are recorded at 1-min intervals and averaged over 30 min; both 1-min recordings and averages are stored on a CR23X datalogger (Campbell Scientific Inc., Logan, Utah, USA).

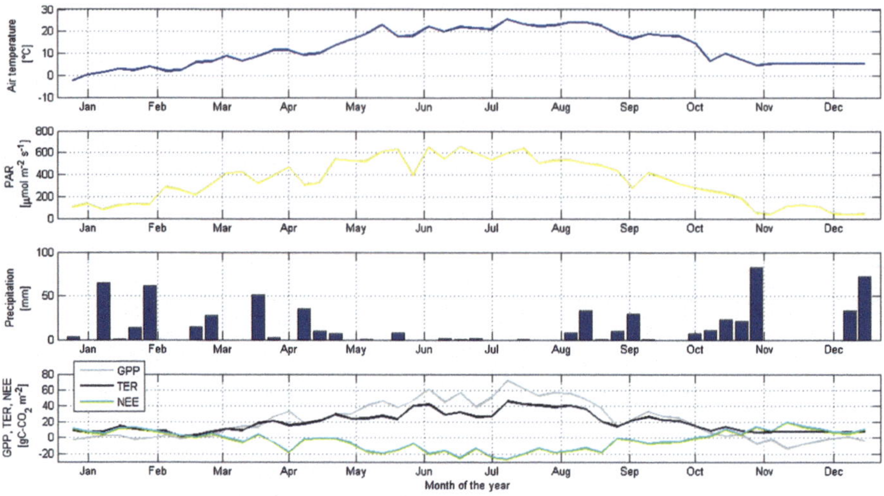

Fig. 11.4 Average air temperature, photosynthetic active radiation (PAR), cumulative precipitation, NEE, GPP and TER in the Mezzolombardo site

Measured raw data were processed using the EdiRe software (R. Clement, University of Edinburgh, UK) to obtain energy and mass fluxes. Flux computation was performed following standard procedures (Aubinet et al. 2000), with data quality control according to Mauder and Foken (2011). Gaps in the dataset were filled using Falge et al. (2001a,b) and Reichstein et al. (2005) methodology.

The average air temperature value for 2009 was 13.5 °C, ranging from −4.5 (January 04) to 26.1 °C (August 18) (Fig. 11.4). Precipitation was mainly distributed during winter and autumn with a total amount of 737.7 mm. The 2009 was characterized by high temperature and low precipitation, in particular during the growing season, in comparison with the 30-years average.

The growing season starts in the middle of March and lasts until the end of September with maximum carbon dioxide uptake at the end of April (because of high radiation and water availability) and in July. In this period, the ecosystem adsorbs more than 200 g C m^{-2} (Fig. 11.4). During the winter season (from October to March the vineyard acted as a C source. The radiation, more than temperature and precipitation, seems to be the main factor that drives carbon fluxes in this vineyard ecosystem (Fig. 11.4). At an annual scale, the vineyard was a net C sink, with cumulated NEE value of 145 g C m^{-2} y^{-1} (Fig. 11.5).

11.1.3 Serdiana (Cagliari) Site

The experimental site is located within the Company Argiolas Winery near Serdiana, Cagliari, Italy (39°21′43″N, 9°07′26″E, 112 m asl) (Fig. 11.1). The

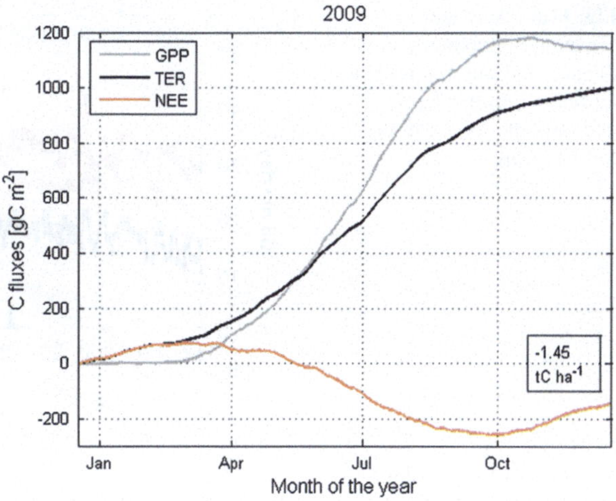

Fig. 11.5 Annual cumulative values of NEE, GPP and TER at the Mezzolombardo site

Vermentino variety grapevines, trained in a Guyot system, are oriented in east-west rows with 0.8 m between plants and 2.0 m between rows. The vegetation is about 2.0 m tall with about 50 % ground shading of the clay soil.

A meteorological station was established nearby to measure the main meteorological variables and data were transmitted by remote telecommunication system using a GPRS web protocol. An Eddy Covariance station was set up to continuously monitor energy and mass fluxes from June to December 2009. A sonic anemometer (CSAT3, Campbell Scientific, Logan, UT, USA) and an open-path gas analyzer (Li-7500, Li-cor, Lincoln, NE, USA) were used to measure vertical wind speed and water vapour and CO_2 concentration, respectively. A 4-component net radiometer (MR40, Eko Instruments, Tokyo, Japan) was used to measure the radiation balance. Eddy Covariance data were acquired at 10 Hz and averaged every 30 min. Four heat plates (HFP01SC, Hukseflux, Delft, NL) were located at a depth of −7 cm, in a transect between rows to have a representative measure of soil heat flux (G). Soil temperature changes in the soil layer above −7 cm were measured to correct for heat storage and estimate G at the surface. In addition, soil temperature and moisture were monitored at −20, −40, and −60 cm in several vineyard locations.

Measured raw data were processed using the EOLO software, which was developed by the University of Sassari according to the FLUXNET network protocol, to obtain energy and mass fluxes. Data quality control was performed following the procedures of Baldocchi et al. (1997), Aubinet et al. (2000), Schmid et al. (2000), and Papale et al. (2006). Gaps in the dataset were filled using Falge et al. (2001a, b) and Reichstein et al. (2005) methodology. During the growing period, the LAI was also monitored by LAI-2000 (Li-cor, Lincoln, NE, USA) based on the principle of interception of the light radiation from vegetation.

The average air temperature value was 19.5 °C, ranging from 5.6 (December 22) to 32.2 °C (July 24) (Fig. 11.6). Precipitation was mainly distributed during fall with a total amount of 143.4 mm. The energy balance closure was used to evaluate

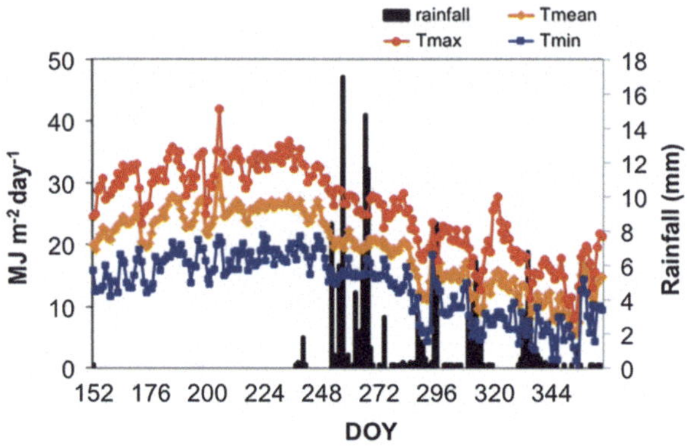

Fig. 11.6 Average (Taverage), maximum (Tmax) and minimum (Tmin) air temperature values, and rainfall amount in the Serdiana site

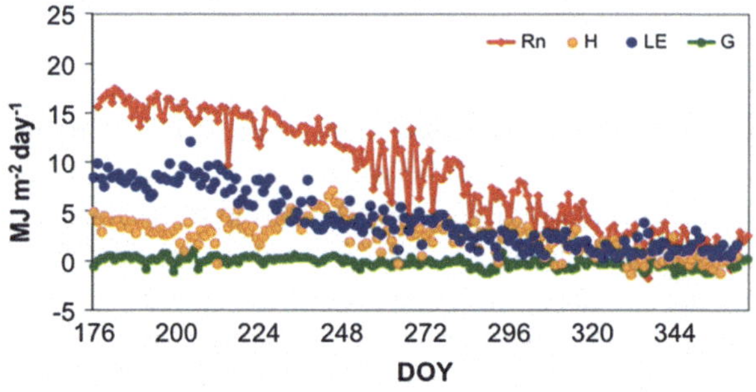

Fig. 11.7 Energy balance components in the Serdiana site

the accuracy of measured data. A discrepancy between 20 and 30 % from the closure is commonly observed in surface energy budget measurements (Wilson et al. 2002), and the closure observed in this research was in line with those reported in the literature (Fig. 11.7). More available energy (R_n–G) was partitioned to latent heat flux (*LE*) than sensible heat flux (*H*) in June and July (*LE* > *H*), presumably because there was adequate water available from irrigation. From late August to the end of the year, energy was approximately equally partitioned between *H* and *LE*.

Carbon exchanges between the ecosystem and the atmosphere were partitioned in NEE, GPP, and ecosystem respiration (Reco). Daily values revealed that NEE was directly affected by high temperature and rainfall (Fig. 11.8a). In particular, a NEE value close to zero was observed in Day Of Year (DOY) 205 when maximum air temperature was about 42 °C. Also, picks in NEE positive values in DOY 253 and 267 were probably due to the soil microbes respiration, which was stimulated by the first rainfall after the summer. Monthly values revealed two periods with different ecophysiological behaviour: the vineyard acted as a C *sink* from June to September (NEE negative values), and C *source* in winter (NEE positive values) (Fig. 11.8b). During the growing period (June-September), the vineyard was able to accumulate about 170 g C, even if the amount varied between months related to weather conditions. From June to the end of the year, the vineyard was a net C *sink*, with cumulated NEE value of 89 g C m^{-2} y^{-1}.

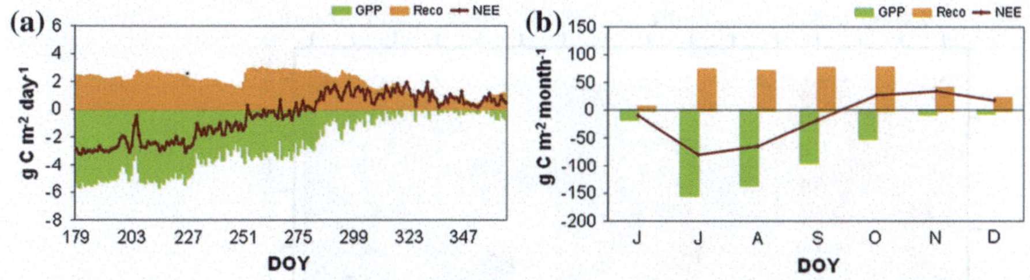

Fig. 11.8 Daily (**a**) and monthly (**b**) values of GPP, NEE, and ecosystem respiration (Reco), in the Serdiana site

11.2 Abandonment and Re-planting Influence on Soil Carbon Stock: The Case of the Sicilian Vineyards

Over the last decades most of the Sicilian vineyards were abandoned or replanted with new varieties thanks to economic incentives by the Sicilian government in order to improve wine quality. While replanting, the soil was often managed with intensive and deep tillage (to prepare the ground for future vineyards), involving changes in SOC stock. In Sicily more than 11,000 ha of vineyard were abandoned and more than 45,000 ha have been subjected to new plantation on a total of 121,000 ha since 2000. Understanding and quantifying the carbon gain and loss in relation to land uses (abandonment and replantation) after old vineyards extirpation is crucial in order to obtain better defined carbon inventories (IPCC 2006). The great diversity of soil types and land use histories will make it difficult to predict the soil GHGs balances at regional or national scale. A representative area in the southwest of Sicily [between the district of Castelvetrano and Campobello di

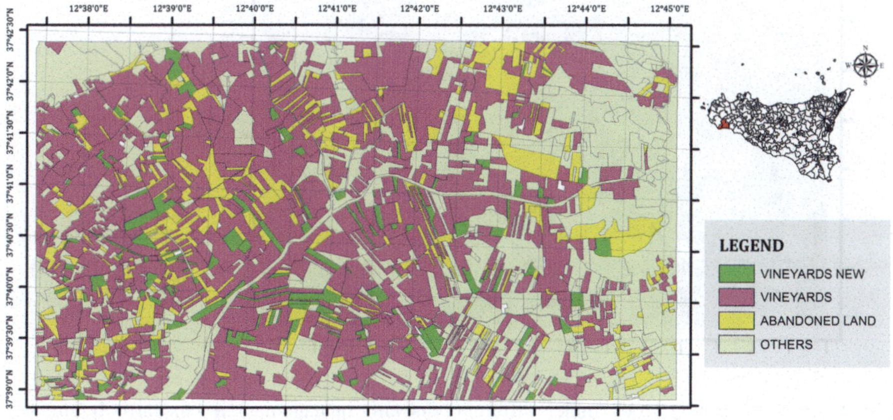

Fig. 11.9 Test area localization and land use of Sicilian study area

Fig. 11.10 Soil carbon stock under different vineyard management **a** vineyard, **b** vineyard replanting, **c** vineyard abandon

Mazara del Vallo (TP)] was selected to study the effect of abandonment or replantation of old vineyard on C stock (Fig. 11.9).

In the last 10 years 10 % of vineyards were abandoned and 37 % were replanted. The SOC stock increased with vineyard abandonment, from 23 to 50 Mg ha^{-1} after 30 years since abandon. On the contrary replanting led to an initial SOC decrease due to intensive tillage during the extirpation of old vineyards and the preparation of field for future vineyard that favour mineralization of SOM and increase erosion loss. After 1 year since replanting, the SOC content was 15 Mg ha^{-1} and it grew up to 25 Mg ha^{-1} after 8 years since replantation (Fig. 11.10).

On average abandonment led to SOC increase of 27 % in comparison to old vineyard, while replanting led to SOC decrease up to 43 %. Considering the average of SOC content for each land use (old vineyard, replanted and abandoned vineyard) and the relative area of land use change during the last 10 years, we estimated a SOC loss in the study area. It was due to high increase of replanted vineyards that was not balanced by abandoned area. In this context, regional land management issues associated to the land use change must be given special attention in order to prevent further losses of organic C from soils, land degradation and potential losses in soil productivity, also considering that vineyard systems response to soil fertility changes vary rapidly, from 1 to 2 years, imposing special consideration to alternative management (Ripoche et al. 2011).

11.3 Conclusions

We present 1 year of carbon and energy balance in three Italian vineyard sites along a north-south transect. All the three sites were net sink of carbon, with a cumulated NEE ranging between very high values (814 g C m^{-2} y^{-1}) in the Negrisia site and 145 g C m^{-2} y^{-1} in the Valle dell'Adige site and 89 g C m^{-2} y^{-1} in the Serdiana site (but considering only fluxes from June to December). Considering the high variability of carbon flux balance in relatively homogeneous ecosystems, such as the three vineyards analysed in this chapter, in our opinion the global distribution of cropland measurement sites is too limited, far from being representative of the huge variety of climate regions, crops and management practices. We also analysed the effect of land use change considering the effect on soil carbon of the abandonment and re-planting of vineyard in Sicily. On average abandonment led to a 27 % SOC increase compared to old vineyards, while replanting led to an initial SOC decrease of up to 43 %. We estimated a SOC loss in the study area during the last 10 years, and we suggest that the regional land management offices must take special care to prevent further losses of organic C from soils.

Acknowledgments The Serdiana site was funded by the Sardinia Region Project SQFVS (Per un Salto di Qualità della Filiera Vitivinicola della Sardegna) Progetto P6 "Centro di ricerca e trasferimento tecnologico nella filiera del vino di qualità", within the Accordo di Programma Quadro in materia di Ricerca Scientifica e Innovazione Tecnologica e realizzato dalle aziende del Convisar–Consorzio Vino e Sardegna: Argiolas, Serdiana, Cantina Gallura, Tempio, Cantina Li Duni, Badesi, Cantina delle Vigne di Piero Mancini, Olbia, Cantina del Mandrolisai, Sorgono, Cantina Trexenta, Senorbì, and by the Italian Ministry of Environment under grant agreement no. 232/2011. We also would like to thanks Mr. Pierpaolo Zara and Angelo Arca (CNR-IBIMET, Sassari) for the assistance and maintenance of the monitoring stations.

References

Allmaras RR, Schomberg HH, Douglas CL Jr, Dao TH (2000) Soil organic carbon sequestration potential of adopting conservation tillage in US croplands. J Soil Water Conserv 55:365–373

Aubinet M, Grelle A, Ibrom A, Rannik U, Moncrieff J, Foken T, Kowalski AS, Martin PH, Berbigier P, Bernhofer Ch, Clement R, Elbers J, Granier A, Grunwald T, Morgenstern K, Pilegaard K, Rebmann C, Snijders W, Valentini R, Vesala T (2000) Estimates of the annual net carbon and water exchange of European forests: the EUROFLUX methodology. Adv Ecol Res 30:113–174

Baldocchi D, Vogel CA, Hall B (1997) Seasonal variations of carbon dioxide exchanges rates above and below a boreal jack pine forest. Agric For Meteor 83:147–170

Desai AR, Bolstad P, Cook BD, Davis KJ, Carey EV (2005) Comparing net ecosystem exchange of carbon dioxide between an old-growth and mature forest in the upper Midwest USA. Agric For Meteor 128(1–2):33–55

Falge E, Baldocchi D, Olson RJ, Anthoni P, Aubinet M, Bernhofer C, Burba G, Ceulemans R, Clement R, Dolman H, Granier A, Gross P, Grünwald T, Hollinger D, Jensen NO, Katul G, Keronen P, Kowalski A, Lai CT, Law BE, Meyers T, Moncrieff J, Moors E, Munger JW, Pilegaard K, Rannik Ü, Rebmann C, Suyker A, Tenhunen J, Tu K, Verma S, Vesala T, Wilson K, Wofsy S (2001a) Gap filling strategies for defensible annual sums of net ecosystem exchange. Agric For Meteor 107:43–69

Falge E, Baldocchi D, Olson RJ, Anthoni P, Aubinet M, Bernhofer C, Burba G, Ceulemans R, Clement R, Dolman H, Granier A, Gross P, Grunwald T, Hollinger D, Jensen NO, Katul G, Keronen P, Kowalski A, Lai CT, Law BE, Meyers T, Moncrieff J, Moors E, Suyker A, Tenhunen J, Tu K, Verma S, Vesala T, Wilson K, Wofsy S (2001b) Gap filling strategies for long term energy flux data sets. Agric For Meteor 107:71–77

Foken T, Wichura B (1996) Tools for quality assessment of surface-based flux measurements. Agric For Meteor 78:83–105

Grove AT (1996) The historical context: before 1850. In: Brandt CJ, Thornes JB (eds) Mediterranean desertification and land use. Wiley, Chichester, UK, pp 13–28

IPCC (2001) Climate Change 2001: Synthesis Report. A Contribution of Working Groups I, II, and III to the Third Assessment Report of the Integovernmental Panel on Climate Change. In: Watson RT, the Core Writing Team (eds) Cambridge University Press, Cambridge, United Kingdom, and New York, NY, USA, pp 398

IPCC (2006) Guidelines for National Greenhouse Gas Inventories, Prepared by the National Greenhouse Gas Inventories Programme. In: Eggleston HS, Buendia L, Miwa K, Ngara T, Tanabe K (eds) IGES, Japan

Margaris NS, Koutsidou E, Giourga C (1996) Changes in traditional Mediterranean land-use systems. In: Brandt CJ, Thornes JB (eds) Mediterranean desertification and land use. Wiley, Chichester, pp 29–42

Mauder M, Foken T (2011) Documentation and instruction manual of the Eddy covariance software package TK3. Univ. Abt, Mikrometeorologie

OIV (2012) Statistical report on world vitiviniculture. International organization of vine and wine. http://www.oiv.int/oiv/info/enstatistiquessecteurvitivinicole#bilan

Papale D, Reichstein M, Canfora E, Aubinet M, Bernhofer C, Longdoz B, Kutsch W, Rambal S, Valentini R, Vesala T, Yakir D (2006) Toward a more harmonized processing of eddy covariance CO_2 fluxes: algorithms and uncertainty estimation. Biogeosci Discuss 3:961–992

Puigdefábregas J, Mendizabal T (1998) Perspectives on desertification: western Mediterranean. J Arid Environ 39:209–224

Reichstein M, Falge E, Baldocchi D, Papale D, Aubinet M, Berbigier P, Bernhofer C, Buchmann N, Gilmanov T, Granier A, Grunwald T, Havrankova K, Ilvesniemi H, Janous D, Knohl A, Laurila T, Lohila A, Loustau D, Matteucci G, Meyers T, Miglietta F, Ourcival JM, Pumpanen J, Rambal S, Rotenberg E, Sanz M, Tenhunen J, Seufert G, Vaccari F, Vesala T, Yakir D, Valentini R (2005) On the separation of net ecosystem exchange into assimilation and ecosystem respiration: review and improved algorithm. GCB 11:1424–1439

Ripoche A, Metay A, Celette F, Gary C (2011) Changing the soil surface management in vineyards: immediate and delayed effects on the growth and yield of grapevine. Plant Soil 339:259–271

Rounsevell MDA, Audsley E, Mortimer D (2002) The impact of the common agricultural policy on land use in Europe. In: Verheye WH (ed) Land cover and land use. Encyclopedia of life support systems (EOLSS), developed under the auspices of the UNESCO. EOLSS Publishers, Oxford

Schmid HP, Grimmond CSB, Cropley F, Offerle B, Hong-Bing Su (2000) Measurements of CO_2 and energy fluxes over a mixed hardwood forest in the mid-western United States. Agric For Meteor 103:357–374

Thornes JB (1995) Mediterranean desertification and the vegetation cover. In: Fantechi R, Peter D, Balabanis P, Rubio JL (eds) Desertification in an European context: physical and socioeconomic aspects. European Commission, Brussels, pp 169–194

Wainwright J (1994) Anthropogenic factors in the degradation of semi-arid lands: a prehistoric case study in southern France. In: Millington AC, Pye K (eds) Environmental change in dry lands: biogeochimical and geomorphological perspectives, Wiley, London, pp 427–441

Watson RT, Noble IR, Bolin B, Ravindramath NH, Verardo DJ, Dokken DJ (2000) Land use, land-use change, and forestry (a special report of the IPCC). Cambridge University Press, Cambridge

Wilson K, Goldstein A, Falge E, Aubinet M, Baldocchi D, Berbigier P, Bernhofer C, Ceulemans R, Dolman H, Field C, Grelle A, Ibrom A, Law BE, Kowalski A, Meyers T, Moncrieff J, Monson R, Oechel W, Tenhunen J, Valentini R, Verma S (2002) Energy balance closure at FLUXNET sites. Agric For Meteor 113:223–243

Chapter 12
Afforestation and Reforestation: The Sicilian Case Study

Juliane Rühl, Luciano Gristina, Tommaso La Mantia, Agata Novara and Salvatore Pasta

Abstract In some regions of the world such as the Northern Hemisphere, the abandonment of agricultural land is one of the most widespread forms of land use change. In general, abandonment is followed by colonization by herbaceous and woody plants. Since the 1950s, wide areas of Southern Italy have been afforested for soil conservation improvement. In order to quantify the effects of agricultural abandonment and artificial afforestation on soil organic carbon (SOC), a dataset of 48 Sicilian sites has been analyzed. Because of their high environmental variability, these sites can be considered as representative of Southern Italy and in general of the Mediterranean basin. Soil samples were taken throughout all bioclimates in different successional stages (cultivated areas: orchards, cereal crops, herb-dominated plant communities, grasslands dominated by perennial grasses, garrigues and low shrublands, maquis, natural forests and in nearby artificially afforested sites (Pine plantations)). The study confirmed that SOC accumulation after agricultural abandonment depends on bioclimate: the highest SOC accumulation was recorded in the meso-mediterranean bioclimate, intermediate in the thermo-mediterranean, and the lowest in the supra-mediterranean bioclimate. Data showed that for C sequestration in the soil, artificial afforestation is not convenient in comparison to natural afforestation by spontaneous secondary succession processes.

12.1 Introduction

In the Mediterranean region, lands cultivated since ancient times are today subject to two contrasting trends: intensification or abandonment. Since the 1950s, vast areas of Southern Italy have been afforested with the aim of soil conservation and

J. Rühl · L. Gristina · T. La Mantia (✉) · A. Novara · S. Pasta
Agricultural and Forest Sciences Department, University of Palermo,
Viale delle Scienze, Ed. 4, Ingr. H, 90128 Palermo, Italy
e-mail: tommaso.lamantia@unipa.it

R. Valentini and F. Miglietta (eds.), *The Greenhouse Gas Balance of Italy*,
Environmental Science and Engineering, DOI 10.1007/978-3-642-32424-6_12

as a reaction to the progressive land abandonment caused by rural depopulation. The objective of this chapter is to present the results of a case study which quantifies the impact of agricultural abandonment and afforestation on soil organic carbon stock. The study has been carried out in Sicily, the southernmost region of Italy, which is characterized by a multitude of environmental sub-regions due to a high variability of local environmental factors such as altitude, outcropping rock types, etc.. For example, in Sicily there is a wide bioclimatic gradient, ranging from the infra-mediterranean to the cryo-oromediterranean thermotype. Therefore, the case study of Sicily may be considered as representative also for other regions of S Italy and of the Mediterranean Basin.

In this chapter, the term "natural afforestation" indicates the spontaneous process of secondary succession that follows agricultural abandonment, and "artificial afforestation" indicates (FAO) the ≪"Man-made Forest": A forest crop raised artificially, either by sowing or planting≫.

12.1.1 Natural Afforestation

Land-use change (LUC) is considered the second greatest cause of carbon (C) emissions after fuel consumption (Watson et al. 2000). During the past 250 years, about 200 Pg of C were released to the atmosphere as a consequence of LUC (Scholes and Noble 2001). Crop cultivation in areas previously covered by native woody vegetation induces huge C losses from biomass and soil and it is today considered, along with the deforestation of tropical areas, one of the main causes of CO_2 concentration increase in the atmosphere.

On the other hand, in some regions of the world, such as the mid-latitudes of the Northern Hemisphere, one of the most important types of LUC is the abandonment of agricultural land. In Italy, between 1961 and 2009, almost 39.2 % of the "Arable Land and Permanent Crops", corresponding to an area of 61,230 km^2, have been abandoned (FAO database).

In general, the abandonment of cultivated or grazed lands is the starting point for secondary succession processes, i.e. the spontaneous colonization of the areas by herbaceous and woody plants. The species composition of the successional plant communities and their turnover rates depend on a variety of abiotic and biotic factors, such as macro- and mesoclimate. In most cases and in the absence of strong disturbance factors such as grazing or wildfires, secondary succession leads within few decades to the creation of forest communities. In the Mediterranean area, the recent abandonment of marginal agricultural areas (pasture and/or arable lands) has caused an increase in the area occupied by pre-forest and forest communities (Bonet 2004). In Italy, total forest area has increased by 17.5 % during the period 1950–1990 (Italian National Statistical Bureau).

Changes in soil organic carbon (SOC) content after agricultural abandonment have been quantified by various authors, and some of them reported an increase in SOC during woody plant encroachment within old fields or grasslands (Prévosto et al. 2006; La Mantia et al. 2007; Montané et al. 2007; Alberti et al. 2011), while others found a decrease in SOC during secondary succession (Paul et al. 2002; Guo and Gifford 2002; Jackson et al. 2002; Goodale and Davidson 2002; Alberti et al. 2008). So, there is a need of further studies on the factors that determine if the soils of the areas subject to secondary succession are either C sinks or C sources. On a regional scale, such kind of information could be particularly important for identifying land-use classes that act as C sinks.

One of the factors that influence SOC dynamics after land abandonment is climate (Jackson et al. 2002). In Italy, a negative relationship between changes in SOC after agricultural abandonment and total annual rainfall has been found, possibly related to soil N stock dynamics (Alberti et al. 2011). This relationship has recently been confirmed for Sicily (La Mantia et al. 2013).

12.1.2 Artificial Afforestation

In Mediterranean countries, the surface of tree plantations has strongly increased during the 20th century. Under semi-arid climatic conditions, afforestation activities have been carried out mostly by using various species of *Pinus* and *Eucalyptus*, and they were often planted in abandoned agricultural or degraded lands. After the Second World War, the surface of afforestations and reforestations increased of 435,906 ha in Italy and of 73,228 ha in Sicily, and areas dedicated to “arboriculture for timber” increased of 122,252 ha in Italy and of 1,137 ha in Sicily (INFC 2007).

Within the last decade, an increasing amount of research has focused on the impact of afforestations on soil parameters, and some studies analysed SOC stocks in afforested sites (see Paul et al. 2002). Some authors found a decrease and others an increase of SOC after afforestation with respect to sites characterized by spontaneous pre-forest plant communities (Jackson et al. 2002; Farley et al. 2004; Kelty 2006; Goberna et al. 2007; Vilà et al. 2007; Fernández-Ondoño et al. 2010; Novara et al. 2013). The C increase or decrease after afforestation is, in fact, determined by many factors, including previous land use, climate, soil type and site preparation (Turner and Lambert 2000; Guo and Gifford 2002; Paul et al. 2002).

Considering the high extent of afforested areas in the Mediterranean countries, understanding the ability of afforested soil to sequester C is therefore relevant.

12.2 Study Sites and Data Analysis

In order to quantify the effects of (1) agricultural abandonment and (2) artificial afforestation on SOC, a joined dataset of original data sampled in various sites in Sicily (34 sites) as well as existing data from literature (14 sites), has been analysed. Since previous studies showed that SOC dynamics in secondary succession

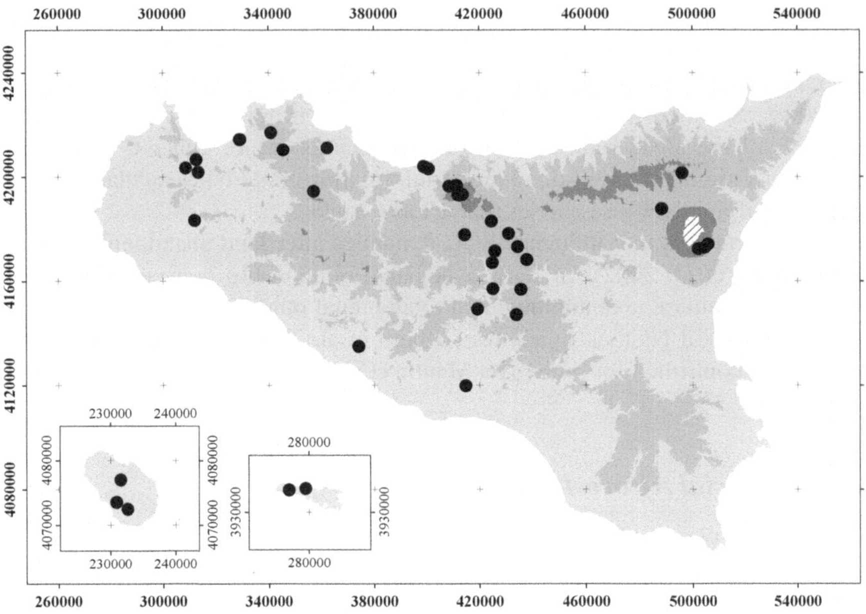

Fig. 12.1 Location of the study sites (*black circles*) within the three most widespread bioclimatic belts (sensu Rivas-Martínez) of Sicily: *light grey* thermo-mediterranean, *medium grey* meso-mediterranean, *dark grey* supra-mediterranean. *Dashed* other bioclimates; climatic data issue from Drago et al. (2002)

areas (Alberti et al. 2011; La Mantia et al. 2013) are influenced by climate, the dataset has been subdivided into three bioclimatic groups of sites (1. thermo-mediterranean, 2. meso-mediterranean, 3. supra-mediterranean), which represent the most widespread bioclimatic belts of Sicily (Fig. 12.1). Thermotype classification follows Rivas-Martínez and Loidi Arregui (1999). The thermotype (It) is a bioclimatic index based on the temperature regime of a site, and it is calculated as follows: $It = (T + M + m) \times 10$, where T is the average annual temperature, M is the average of the maximum temperatures of the coldest month, and m is the average of the minimum temperatures of the coldest month. In total, 16 thermo-, 22 meso-, and 10 supra-mediterranean sites have been analyzed.

12.2.1 Original Sampling Data

Soil sampling was carried out in sites where series of secondary succession, i.e. different progressive successional stages, were present. Throughout all bioclimates, we defined as successional stages the following six land use classes: (1) CU = cultivated areas (orchards, cereal crops); (2) HE = herb-dominated plant communities without presence of woody species (grazed pastures, old fields abandoned by agriculture since a few years); (3) PG = grasslands dominated by perennial grasses; (4) GS = garrigues and low shrublands; (5) MA = maquis; (6) FO = natural forests. In every bioclimate the most representative succession pathways were chosen, for each site, except for supra-mediterranean belt, where the successional stages 3 (PG) and 5 (MA) are missing.

In addition to the soil samples taken in different successional stages, also samples in nearby artificially afforested sites (AF) were analyzed. All sampled afforested sites were Pine plantations, wherein the dominant Pine species depended on the bioclimatic belt (*Pinus halepensis* Mill. and *Pinus pinea* L. in the thermo- and meso-mediterranean belt, *Pinus nigra* J.F. Arnold in the supra-mediterranean bioclimate).

Soil was sampled according to the protocol of the Italian National Inventory of Forests and Forest Carbon pools (Gasparini et al. 2008). After litter removal, mineral soil samples were collected at depths of 0–10 and 10–30 cm, using a 340 and a 680-cm^3 cylinder, respectively. For every site and succession stage/afforestation, three replicates of both soil depths were taken, respectively.

The soil samples were gently broken, passed through a 2-mm sieve, air dried, treated with HCl 2:1 to remove carbonates and then analysed for C content using a CHN-Elemental Analyzer. SOC content was first expressed as a percentage (g of C per 100 g of dry soil $\times$ 100) and then converted to tons per hectare based on bulk density (BD) and soil depth. Bulk density was measured based on the volume of the collected sample and the weight of dry soil in the sample (Blake and Hartge 1986). Mass correction of SOC stock estimates based on BD is crucial for estimating the effects of land-use change because land-use change is always accompanied by changes in BD. All data were adjusted to 30 cm depth.

12.2.2 Data from Literature

Existing literature including SOC analyses in Sicilian sites was screened for data taken in successional stages or in land use classes, which could be compared, as similar to stages of secondary succession. Moreover, SOC data of afforested areas (*Pinus* spp., *Eucalyptus* spp.) in proximity to the successional areas were included in the analyses (Ballatore and Fierotti 1970; Raimondi et al. 1983; Panno et al. 1986; Ferro et al. 2008; Agnese et al. 2011; Novara et al. 2012; La Mantia et al. 2013).

12.2.3 Data Analysis

For data analyses, average SOC stock values were computed from sites and replicated for each succession stage and for afforested sites within the three bioclimatic groups.

Furthermore, in order to confront SOC among succession stages within each bioclimate, an average index of SOC content change has been used:

Succession Carbon Change Index

$$(\mathrm{SCCI}) = (\mathrm{C}_{\mathrm{older}} - \mathrm{C}_{\mathrm{younger}})/\mathrm{C}_{\mathrm{younger}},$$

where $\mathrm{C}_{\mathrm{older}}$ = SOC content in the older successional stage, and $\mathrm{C}_{\mathrm{younger}}$ = SOC content in the younger successional stage. A positive value indicates a gain in carbon stock from a younger succession stage to an older stage, while a negative value indicates a loss in carbon stock.

In order to evaluate for each bioclimate the effect of artificial afforestation in comparison with a considered stage of succession, another average carbon change index has been used as follows:

Afforestation Carbon Change Index

$$(\mathrm{ACCI}) = (\mathrm{C}_{\mathrm{afforestation}} - \mathrm{C}_{\mathrm{succession\ stage}})/\mathrm{C}_{\mathrm{succession\ stage}},$$

where $\mathrm{C}_{\mathrm{afforestation}}$ = SOC content in the afforestation, and $\mathrm{C}_{\mathrm{succession\ stage}}$ = SOC content in the succession stage. A positive value indicates a gain in carbon stock by afforestation activities in confront with the respective succession stage, while a negative value indicates a loss.

Lastly, the C stock data have been used to calculate the total C gain in areas subject to secondary succession in the region of Sicily between the years 1990 and 2006. Surface data of agricultural abandonment and areas subject to secondary succession processes were elaborated from CORINE land cover classification data.

12.3 Results

12.3.1 SOC Variation in Natural Afforestation

In all bioclimates, SOC stock increased more than twice from cultivation to the last stage of succession (Fig. 12.2). A noteworthy difference in SOC stock between the bioclimates has been observed: from cultivated soils to forest soils, SOC stock increased from 28.31 to 78.60 Mg ha^{-1} in the thermo-mediterranean bioclimate, from 45.51 to 106.29 Mg ha^{-1} in the meso-mediterranean bioclimate and from 31.25 to 73.70 Mg ha^{-1} in the supra-mediterranean bioclimate.

Furthermore, in all bioclimates SOC stock increased from one succession stage to the next one (Fig. 12.3). In the thermo-mediterranean belt, there was a high increase in SOC from the cultivated soils to the soils under herb-dominated plant communities. So, the cessation of soil tillage or chemical weed control after cultivation abandonment seems to be an important factor for SOC accumulation.

On the contrary, we observed only a low SOC stock increase from herb-dominated plant communities to perennial grasslands, and SOC gain during the woody-dominated stages of succession assumed intermediate values.

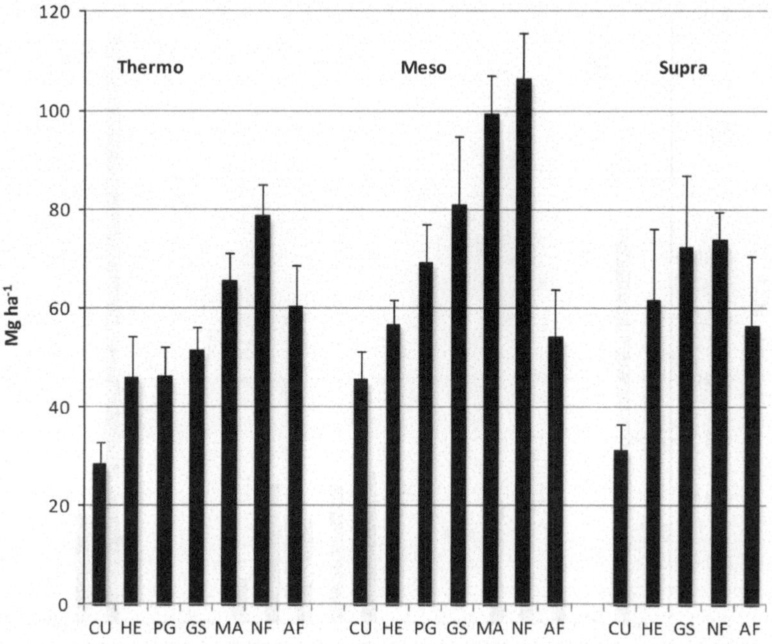

Fig. 12.2 SOC stock (Mg ha^{-1}) in the various stages of secondary succession and in the afforested sites and in three bioclimates (0–30 cm soil depth). Abbreviations: *CU* cultivated areas, *HE* herb-dominated plant communities, *PG* perennial grasslands, *GS* garrigues and low shrublands, *MA* maquis, *NF* natural forests, *AF* artificial afforestations; *Thermo* thermo-mediterranean, *Meso* meso-mediterranean, *Supra* supra-mediterranean bioclimate

In the meso-mediterranean bioclimate, relative SOC stock increase with cultivation abandonment, i.e. from cultivated soils to herb-dominated soils, is only half as much as in the thermo-mediterranean belt. On the other hand, there was a higher increase from herb communities to perennial grasslands. SOC gain in the older succession stages was similar to the thermo-mediterranean belt.

Finally, in the supra-mediterranean belt SOC stock increase was very high with cultivation abandonment: SOC stock doubles from cultivated soils to herb-dominated soils. In contrast, SOC gain in the older succession stages was lower in comparison to the other two bioclimates.

12.3.2 SOC Variation in Artificial Afforestation

Results showed a lower SOC stock in artificially afforested soil in comparison with natural afforested soil (Fig. 12.2). More in detail, in the thermo-mediterranean belt, artificial afforestations showed higher SOC contents with respect to natural afforestations up to the succession stage of garrigues and low shrublands, and

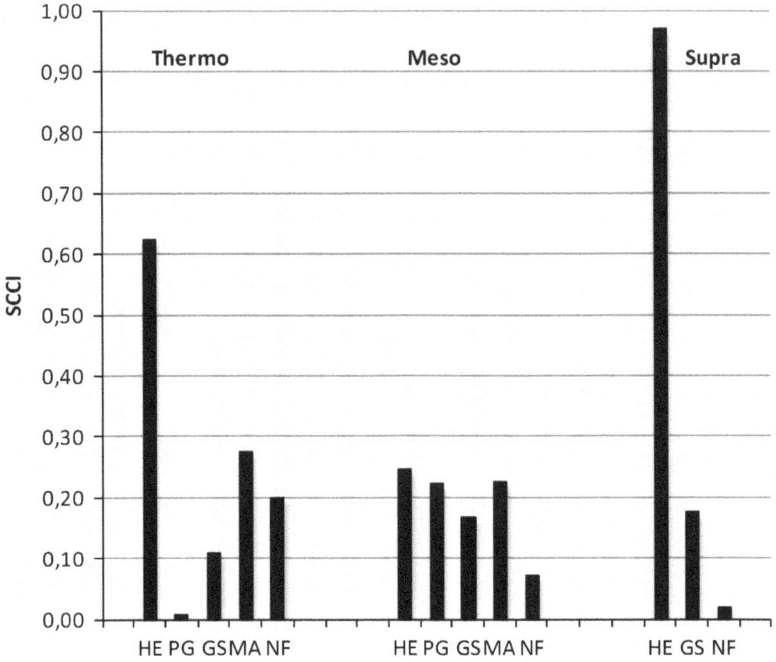

Fig. 12.3 Succession Carbon Change Index (SCCI) in the three bioclimates, indicating SOC stock variation between the various stages of secondary succession. Abbreviations: *CU* cultivated areas, *HE* herb-dominated plant communities, *PG* perennial grasslands, *GS* garrigues and low shrublands, *MA* maquis, *NF* natural forests, *AF* artificial afforestations; *Thermo* thermo-mediterranean, *Meso* meso-mediterranean, *Supra* supra-mediterranean bioclimate

we observed a very high gain in SOC stock in afforested soils with respect to cultivated soils (Fig. 12.4). On the contrary, maquis and natural forests showed higher SOC stocks than afforested sites.

In the meso-mediterranean bioclimate, SOC content gain by artificial afforestation was observed only with respect to cultivated soils, and it was, in addition, quite low, while all other secondary succession stages showed higher SOC stocks than the afforested sites.

Also in the supra-mediterranean bioclimate SOC stock gain by artificial afforestation was found only with respect to cultivated soils. The carbon change index between artificial afforestation and cultivated soils was higher in supra- than in meso-mediterranean conditions.

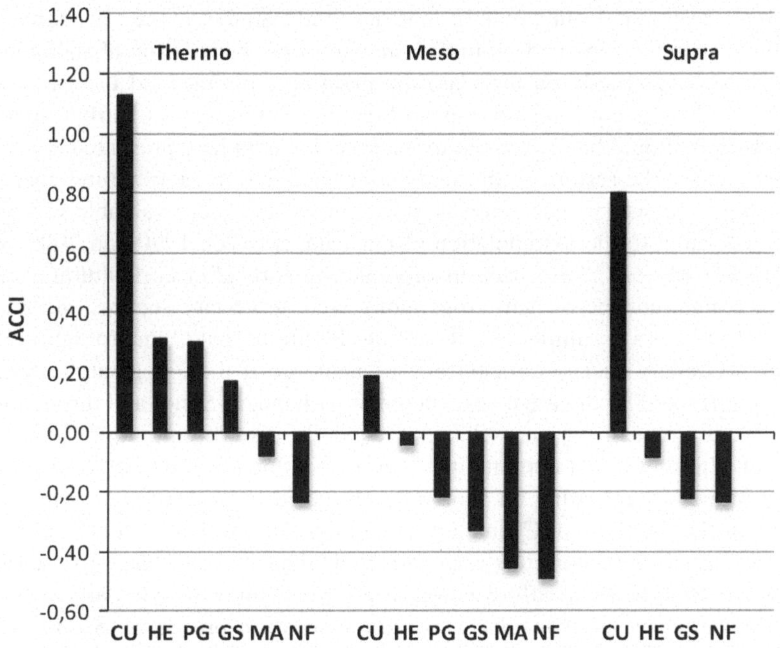

Fig. 12.4 Afforestation Carbon Change Index (ACCI) in the three bioclimates, indicating SOC stock variation between the afforested sites and the various stages of secondary succession, including the cultivated stage. Abbreviations: *CU* cultivated areas, *HE* herb-dominated plant communities, *PG* perennial grasslands, *GS* garrigues and low shrublands, *MA* maquis, *NF* natural forests, *AF* artificial afforestations; *Thermo* thermo-mediterranean, *Meso* meso-mediterranean, *Supra* supra-mediterranean bioclimate

12.4 Conclusions

In the last decades, abandonment and afforestation have strongly modified agricultural systems in Italy. Regarding abandonment, the colonization of abandoned areas by spontaneous vegetation has some important impacts on the environment, for example on water balance, carbon dynamics and biodiversity. The present study confirmed the results of various studies conducted in other biomes, which pointed out that SOC accumulation after agricultural abandonment depends on climate (Alberti et al. 2011; Jackson et al. 2002). SOC accumulation in Sicily differed between the studied bioclimates, indicating the highest total SOC accumulation for the meso-mediterranean belt, intermediate values for the thermo-mediterranean belt, and the lowest values for the supra-mediterranean belt.

It must be pointed out, though, that our study analysed the abandonment of permanent woody crops and ploughed annual crops, but not the abandonment of pastures. Since in Sicily pastures are the main agricultural land use in the areas characterized by a supra-mediterranean bioclimate, research on carbon dynamics in ex-pastures should be carried out to integrate the data here presented.

Furthermore, the results of this study confirmed that the agricultural abandoned areas where secondary succession is going on are the more valuable C sinks. In Sicily, according to the extrapolation of our data, between 1990 and 2006, a total of 6,835,431 Mg of C have been incorporated in soils after agricultural abandonment of annual and permanent crops along with secondary succession processes (Table 12.1). It is very important, though, to highlight that C incorporation in soil through secondary succession can only be achieved if these vegetation dynamics are not interrupted by disturbances. On the other hand, in Sicily and throughout the whole Mediterranean region, wildfires frequently disturb agricultural abandoned areas, thus biasing forest communities' set up. So, the areas where secondary succession processes are going on should be protected from disturbances in order to obtain a further increase of C sink capacity of agriculturally abandoned soils.

Regarding afforestation, this review of SOC data showed that for C sequestration in the soil, artificial afforestation is not convenient in comparison to natural afforestation by spontaneous secondary succession processes. So, in order to increase C storage in abandoned agricultural lands, management choices should aim at the protection of these spontaneous processes, and could also actively enhance spontaneous woody encroachment by bird-mediated seed addition or selective seedling planting techniques. In contrast, in strongly degraded areas or in abandoned agricultural lands where severe disturbance factors such as frequent wildfires and overgrazing occur, afforestation could be a management choice, because under intense and frequent disturbance factors secondary succession

Table 12.1 Gain in soil C (Mg) in areas subject to agricultural abandonment (change from cultivated to abandoned soils) and progressive secondary succession (change from a younger succession stage to an older one) in the three considered bioclimatic belts of Sicily from 1990 to 2006

Change between 1990 and 2006	Thermo	Meso	Supra
From CU to HE/GR	694,075	672,848	1,213
From CU to GS/MA	966,958	1,496,419	15,585
From CU to FO	438,479	732,179	4,013
Total agricultural abandonment	2,099,513	2,901,446	20,811
From HE/GR to GS/MA	73,032	773,696	61,351
From HE/GR to FO	73,980	382,926	19,548
From GS/MA to FO	176,068	250,510	2,549
Total progressive succession	323,081	1,407,132	83,448
Total abandonment + succession	2,422,594	4,308,578	104,260

CU cultivated areas, *HE* herb-dominated plant communities, *PG* perennial grasslands, *GS* garrigues and low shrublands, *MA* maquis, *NF* natural forests, *AF* artificial afforestations; *Thermo* thermo-mediterranean, *Meso* meso-mediterranean, *Supra* supra-mediterranean bioclimate is often blocked in a steady state. As our data analyses showed, however, these kinds of afforestation interventions are effective for SOC stock increase only up to a certain stage of secondary succession, which depends on bioclimate: while under thermo-mediterranean conditions artificial afforestation is convenient in old fields, pastures, perennial grasslands, garrigues and low shrublands, it is not convenient in maquis. In the other two bioclimatic belts, afforestation is convenient only if carried out immediately after agricultural abandonment, but not if secondary succession has already induced the formation of grasslands or woody plant communities. Of course, in afforestations where biomass production and not C accumulation is the major objective, such as in short rotation forestry areas, the complex carbon balance must be carefully evaluated.

In conclusion, the considerations made above could be used to better address future management policies on agriculturally abandoned areas.

References

Agnese C, Bagarello V, Baiamonte G, Iovino M (2011) Comparing physical quality of forest and pasture soils in a Sicilian watershed. For Range Wildland Soils 75(5):1958–1970

Alberti G, Leronni V, Piazzi M, Petrella F, Cairota P, Peressotti A, Piussi P, Valentini R, Cristina L, La Mantia T, Novara A, Rühl J (2011) Impact of woody encroachment on soil organic carbon and nitrogen in abandoned agricultural lands along a rainfall gradient in Italy. Reg Environ Change 11(4):917–924

Alberti G, Peressotti A, Piussi P, Zerbi G (2008) Forest ecosystem carbon accumulation during a secondary succession on Eastern Prealps (Italy). Forestry 81:1–11

Ballatore GP, Fierotti G (1970) Primo contributo allo studio dei Monti "Le Madonie". Quaderni di Agronomia 6:157–214

Blake GR, Hartge KH (1986) Bulk density. In: Klute A (ed) Methods of soil analysis, Part1, 2nd edn. Agronomy Monograph, Vol 9. American Society of Agronomy, Madison, WI:363–375

Bonet A (2004) Secondary succession of semi-arid Mediterranean old-fields in south-eastern Spain: insights for conservation and restoration of degraded lands. J Arid Environ 56:213–233

Drago A, Lo Bianco B, Monterosso I (2002) Atlante climatologico della Sicilia. Assessorato Agricoltura e Foreste. Servizio Informativo Agrometereologico Siciliano. Regione Siciliana. www.sias.regione.sicilia.it

FAO Database -http://faostat.fao.org/site/377/default.aspx#ancor. Last access Nov 2011

Farley KA, Kelly EF, Hofstede RGM (2004) Soil organic carbon and water retention following conversion of grassland to pine plantation in the Ecuadorian Andes. Ecosystems 7:729–739

Fernández-Ondoño E, Serrano LR, Jiménez MN, Navarro FB, Díez B, Martín F, Fernández J, Martínez FJ, Roca A, Aguilar J (2010) Afforestation improves soil fertility in south-eastern Spain. Eur J For Res 129:707–717

Ferro V, Bagarello V, Birtone M, di Stefano C, Iovino M, Lo Papa G, Panno M, Pomilla S (2008) Atlante della qualità dei suoli del bacino del fiume Imera meridionale. Edizioni Fotograf, pp 360

Gasparini P, di Cosmo L, Morelli S, Paletto A, Tosi V, Huez B, Gregori E, Rodeghiero M, Frizzera L, Tonolli S (2008) Manuale di campagna per il rilievo degli attributi integrativi (Fase 3+) INFC

Goberna M, Sánchez J, Pascual JA, García C (2007) *Pinus halepensis* Mill. plantations did not restore organic carbon, microbial biomass and activity levels in a semi-arid Mediterranean soil. Appl Soil Ecol 36:107–115

Goodale CL, Davidson EA (2002) Uncertain sinks in the shrubs. Nature 418:593–594

Guo LB, Gifford RM (2002) Soil carbon stocks and land use change: a meta-analysis. Glob Change Biol 8(4):345–360

INFC (2007) Le stime di superficie 2005. Inventario Nazionale delle Foreste e dei Serbatoi Forestali di Carbonio. MiPAF—Ispettorato Generale Corpo Forestale dello Stato, CRA—ISAFA, Trento. http://www.infc.it

Jackson RB, Banner JL, Jobbagy EG, Pockman WT, Walls DH (2002) Ecosystem carbon loss with weedy plant invasion of grassland. Nature 418:623–626

Kelty MJ (2006) The role of species mixtures in plantation forestry. For Ecol Manage 233:195–204

La Mantia T, Oddo G, Rühl J, Furnari G, Scalenghe R (2007) Variazione degli stock di carbonio in seguito ai processi di abbandono dei coltivi: il caso studio dell'isola di Pantelleria (Tp). Foresta@ 4(1):102–109

La Mantia T, Gristina L, Rivaldo E, Pasta S, Novara A, Rühl J (2013) The effects of post-pasture woody plant colonization on soil carbon and above ground litter carbon and nitrogen along a bioclimatic gradient. iForest 6:238–246

Montané F, Rovira P, Casals P (2007) Shrub encroachment into mesic mountain grasslands in the Iberian Peninsula: effects of plant quality and temperature on soil C and N stocks. Global Biogeochemical Cycles 21:GB4016, doi:10.1029/2006GB002853

Novara A, La Mantia T, Barbera V, Gristina L (2012) Paired-site approach for studying soil organic carbon dynamics in a Mediterranean semiarid environment. Catena 89(1):1–7

Novara A, Gristina L, La Mantia T, Rühl J (2013) Carbon dynamics of soil organic matter in bulk soil and aggregate fraction during secondary succession in a Mediterranean environment. Geoderma 193–194:213–221

Panno M, Bazan E, Petronici C (1986) La materia organica de los suelos composición acídica de "lopidos" en suelos de bosque. Ciencia del suelo 2:155–159

Paul KI, Polglase PJ, Nyakuengama JG, Khanna PK (2002) Change in soil carbon following afforestation. For Ecol Manage 168:241–257

Prévosto B, Dambrine E, Coquillard P, Robert A (2006) Broom (*Cytisus scoparius*) colonization after grazing abandonment in the French Massif Central: impact on vegetation composition and resource availability. Acta Oecol 30:258–268

Raimondi S, Dazzi C, Cirrito V (1983) Modello di studio integrato del territorio (Ficuzza-Palermo). Nota n. 5—I suoli. In: AA.VV. Modello di studio integrato del territorio (Ficuzza-Palermo). Quaderno di Agronomia 10:89–130

Rivas-Martínez S, Loidi Arregui L (1999) Bioclimatology of the Iberian Peninsula. In: Rivas-Martínez S, Fernández-Gonzáles F, Loidi J (eds) Checklist of plant communities of Iberian Peninsula, Balearic and Canary Islands to suballiance level. Itinera Geobotanica, vol 13, pp 353–451

Scholes RJ, Noble IR (2001) Storing carbon on land. Science 294:1012–1013

Turner J, Lambert MJ (2000) Change in organic carbon in forest plantation soils in eastern Australia. For Ecol Manage 133:231–247

Vilà M, Vayreda J, Comas L, Ibañez JJ, Mata T, Obón B (2007) Species richness and wood production: a positive association in Mediterranean forests. Ecol Lett 10:241–250

Watson RT, Noble IR, Bolin B, Ravindramath NH, Verardo DJ, Dokken DJ (2000) Land use, land-use change, and forestry (a special report of the IPCC). Cambridge University Press, Cambridge

Chapter 13
Afforestation and Reforestation: The Friuli Venezia Giulia Case Study

Giorgio Alberti, Gemini Delle Vedove, Silvia Stefanelli and Giuseppe Vanone

Abstract In the last decades, the European Union has favored the conversion of marginal croplands to forest plantations also to mitigate climate change by increasing carbon storage in the biosphere. In Italy, recent estimates report that forest plantations cover a total area of 122,252 ha. The aim of the present paper was to quantify carbon stock and annual carbon flux in mixed broadleaf and poplar plantation in Friuli Venezia Giulia plain. Overall, 2,592 ha of mixed broadleaf plantation (1992–2006) and 3,113 ha of poplar plantations (2002–2006) have been established in the Region with a standing carbon stock of 223,400 and 397,753 Mg C, respectively. Soil is the largest carbon pool in both plantation types (60 and 63 % of total ecosystem carbon stock). The total annual carbon sequestration is respectively 8,855 and 40,286 Mg C y^{-1} whereas soil accounts for 13 % in broadleaf plantations and 9 % in poplar plantations.

13.1 Introduction

Since the origin of agriculture, human population and its consumption of resources have increased and forest and other natural areas have been gradually turned into farmlands, pasturelands and cities. In fact landscape is the result of an alternation of periods of forest expansion (i.e. succession) and periods of reduction in forest areas.

In the last decades, the European Union has supported the conversion of marginal croplands to forest plantations in order to achieve several targets: to diversify

G. Alberti (✉) · G.D. Vedove
Department of Agricultural and Environmental Sciences, University of Udine, via delle scienze 206, 33100 Udine, Italy
e-mail: giorgio.alberti@uniud.it

S. Stefanelli · G. Vanone
Friuli Venezia Giulia Region, Central Direction of Agricultural, Natural and Forest Resources, Friuli Venezia Giulia, Italy

R. Valentini and F. Miglietta (eds.), *The Greenhouse Gas Balance of Italy*, Environmental Science and Engineering, DOI 10.1007/978-3-642-32424-6_13

farmers' income through high quality wood production in the plain, to support agricultural products' prices by reducing cropland area; to diversify the landscape by restoring some natural elements; to mitigate climate change by increasing carbon storage in the biosphere. With reference to the latter aspect, the net carbon exchange in terrestrial ecosystems is the result of a delicate balance between uptake and emissions which show a diurnal, seasonal and annual variability. Consequently, small changes in the relative rates of carbon uptake through photosynthesis and/or carbon loss through respiration can have profound effects on the accumulation of CO_2 in the atmosphere. The Kyoto protocol suggests that management of natural terrestrial carbon sinks, primarily afforestation and reforestation at a global scale, can increase the sink strength and reduce atmospheric CO_2 concentration.

The terms used in this chapter refer to those adopted by the IPCC Guidelines (2000) which define afforestation as the "planting of new forests on lands which, historically, have not contained forests" and reforestation as "the establishment of trees on land that has been cleared of forest within the relatively recent past". The distribution of sources and sinks of carbon over the land surface is dominated by changes in land use: in the tropics current rates of deforestation are responsible for a large loss of carbon; in the north mid-latitudes past changes in land use (i.e. natural forest recovery, forest plantations) can explain much of the observed carbon sink (Houghton and Hackler 2000; Pacala et al. 2001; Houghton 2003).

In Italy, recent estimates report that forest plantations cover a total area of 122,252 ha (54 % poplar plantations; 34 % mixed broadleaf plantations; 12 % coniferous plantations) (Gasparini and Tabacchi 2011). This area is only 1.4 % of total Italian forest area and covers only 0.4 % of country surface, but it has been increasing year by year in the last decades following new European Union regulations and incentives. Thus, the role of these plantations for carbon sequestration has become quite relevant (Magnani et al. 2005) as afforestation and reforestation do not only increase above-ground carbon stocks, but also restore and increase carbon into soil. At steady state, each soil has a theoretical equilibrium carbon content depending on the nature of vegetation, precipitation and temperature (Paul et al. 2002; Guo and Gifford 2002). This equilibrium is the result of the balance between inflows and outflows of the pool and can be disturbed by land use change until a new equilibrium is reached in the new ecosystem. During land use changes, soil can act as a source or a sink of carbon according to the ratio between inflows and outflows. Rodeghiero (2006) reports a carbon accumulation of 0.33 Mg C ha^{-1} y^{-1} following conversion of a cropland into a forest plantation. Guo and Gifford (2002), in their comprehensive meta-analysis, report an increase of 18 % in soil carbon stocks after cropland afforestation.

The Friuli Venezia Giulia Region, following the application of the measure 06 of the EU 2080/92 Regulation and the application the Rural Development Plan Rural 2000–2006 (measure h and i.1), encouraged the conversion of agricultural and non-agricultural lands into forest plantations.

The aim of the present study can be described as follows (i) to quantify total carbon stock and its partitioning among different ecosystem pools (i.e.

aboveground, roots and soil) and (ii) to quantify annual carbon sequestration among these different pools both in mixed broadleaf and poplar plantations in the Friuli Venezia Giulia Region. To achieve these two objectives, a chronosequence approach was adopted.

13.2 Study Area and Data Analysis

The study area is the Friuli Venezia Giulia plain where climatic conditions are quite uniform (average annual temperature: 13.5 °C; average annual precipitation: 1,100–1,500 mm). However, depending on the substrate characteristics, this area can be divided into two sub-areas: the Northern part is characterized by a very permeable substrate made up of pebbles and gravel whose diameter decreases moving towards the valley; the Southern part is characterized by clay soils with an high water retention potential. In particular, this last area is quite suitable especially for poplar cultivation because of the high water requirements of this species.

The basic unit considered in this work coincided with individual land parcels or sets of contiguous parcels uniform in terms of stand characteristics (i.e. species composition, structure, age). This study considered mixed broadleaf plantations planted after 1992 with a rotation period of at least 20 years. Moreover, poplar plantations realized after 2002 with a rotation period of 8–10 years were also considered. On the basis of the data given by the Regional Forest Administration, a set of 36 mixed broadleaf plantations and 9 poplar plantations distributed across the plain and across different age classes was selected.

Within each stand, 5 points within a 50 × 50 m grid were randomly selected avoiding plantation's edge. At each point, a 10 m radius plot (314 m^2) was set up for dendrometric measurements. In each plot, all standing trees were recognized, numbered, diameters at breast height (DBH, 1.30 m) and total height were measured. Total aboveground carbon stock was determined according to specific allometric equations for broadleaf plantations (Alberti et al. 2006).

Instead, specific allometric equations elaborated by the Regional Forest Administration for poplar were used. Root biomass was derived using a root/shoot ratio of 0.26 (IPCC 2003).

Aboveground NPP (Mg ha^{-1} y^{-1}) was modeled assuming a logistic growth model. In particular, it was considered equal to the first derivative of Richard's logistic model (Cooper 1983; Hooker and Compton 2003):

$$B = B_{\max} \cdot \left[1 - \exp(a \cdot t)\right]^{b}$$

where B_{max} is the maximum aboveground carbon stock (Mg C ha^{-1}), t is stand age (years), a and b are fitting parameters.

For the mixed broadleaf plantations, a subsample of six stands distributed in two age classes (10 and 20 years) were considered for soil carbon stock assessment. Instead, in the case of poplar plantation, all previously nine selected stands were considered (3, 6 and 9 years). Moreover, adjacent cropland (i.e. maize and wheat for mixed broadleaf and poplar plantations, respectively) were also included in the analysis. Within each forest stand or agricultural crop, five points were randomly selected. Then eight soil cores around each point were collected at two depths (0–15 and 15–30 cm) after litter removal. Prior to processing, samples were kept at 4 °C. Each sample was mixed and sieved through a 2-mm sieve and then dried at 102 °C for 48 h. Dried material was ground to a fine powder, treated with HCl 2:1 to remove carbonates and then analyzed for organic C and N using a CHN Elemental Analyzer. Soil bulk density for each depth was estimated at each selected point following the excavation method (Elliott et al. 1999). Measured values were then plotted against aboveground carbon stocks and linear regressions were used to quantify soil carbon stocks and soil carbon stock changes at different stand ages.

13.3 Results for the Friuli Venezia Giulia Case Study

By using the data provided by the Regional Forest Administration, it was possible to determine the annual cropland surface converted to mixed broadleaf and poplar plantations (Fig. 13.1). In this last case, only plantations realized after 2002 were considered as at the end of the rotation period (8–10 years) the stand is usually clear-cut. Thus, our carbon budget considered the normal rotation period length for both plantation types. The annual conversion rate was between 3 and 532 $ha^{-1}\ y^{-1}$ and between 238 and 606 $ha\ y^{-1}$ for mixed broadleaf and poplar plantations, respectively. Overall 2,592 ha of mixed broadleaf plantation

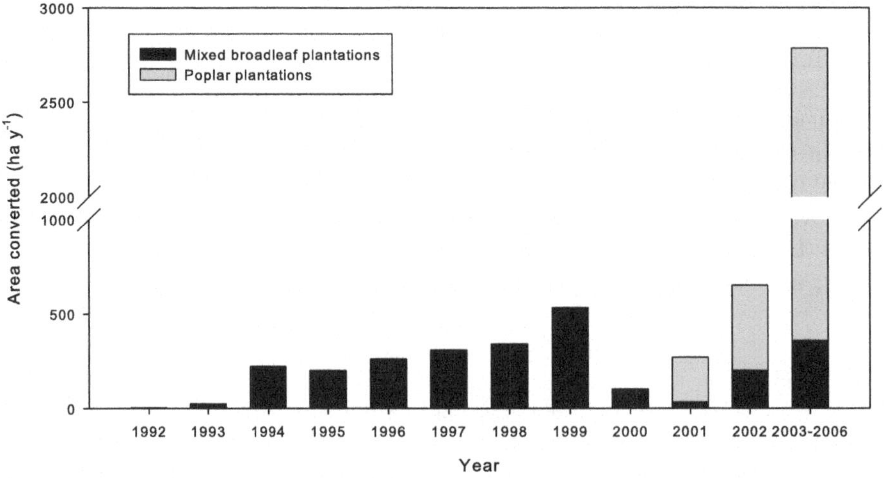

Fig. 13.1 Annual area converted into mixed broadleaf (*black*) and poplar plantations (*grey*) in Friuli Venezia Giulia (Italy)

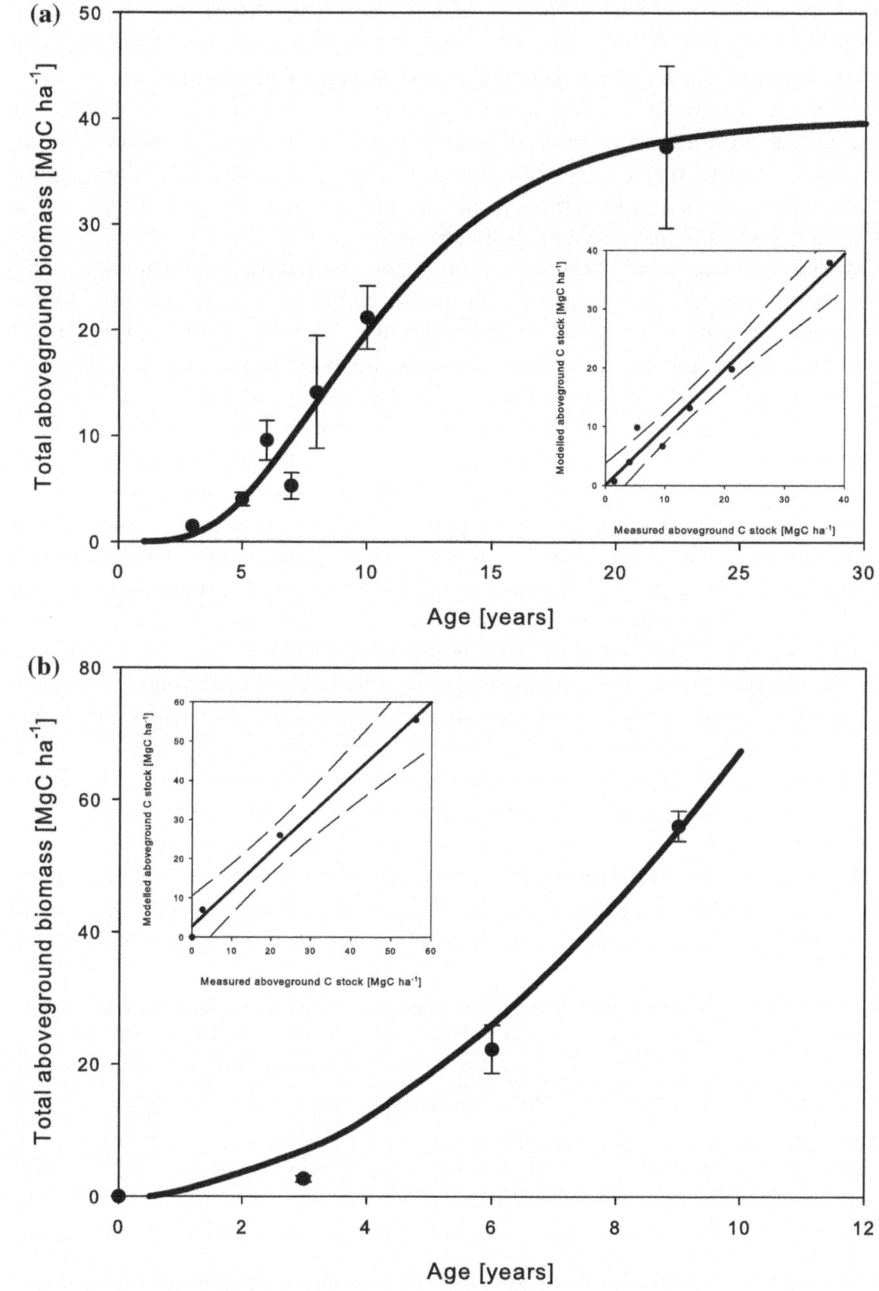

Fig. 13.2 Modeled aboveground carbon stock across the broadleaf plantation chronosequence (**a**) and poplar chronosequence (**b**) as a function of stand age. *Dots* represent measured data, *continuous line* represents modeled data. *Vertical bars* are standard errors. Correlation between modeled and measured data is reported in the inserted panel with 95 % confidence intervals (broadleaf: $Y = 0.99\ X + 0.026$, $R^2 = 0.97$, $p < 0.001$, intercept is not significantly different from zero; poplar: $Y = 0.96\ X + 2.73$, $R^2 = 0.99$, $p = 0.004$, intercept is not significantly different from zero)

(1992–2006) and 3,113 ha of poplar plantations (2002–2006) have been established in the Region.

The average density per hectare in mixed broadleaf plantations was equal to 1,690 ± 29 stumps ha^{-1}. The species composition was quite varied although there was a prevalence of common ash (*Fraxinus excelsior* L.), cherry (*Prunus avium* L.), maple (*Acer campestre* L.) and oak (*Quercus robur* L.). Aboveground carbon stock (stems and branches) in the selected plantations ranged from 1.5 to 37.3 Mg C ha^{-1} at 3 and 23 years, respectively.

Modeled aboveground carbon stock grew asymptotically with stand age, and a good correlation between modeled and measured values was found (Fig. 13.2a). On average, mean aboveground carbon stock was 25.3 Mg C ha^{-1} (Table 13.1). According to the model, maximum aboveground NPP is reached at 9–10 years (3.4 Mg C ha^{-1} y^{-1}) and then it gradually decreases to 0.4 Mg C ha^{-1} y^{-1} at 23 years. Aboveground productivity is therefore rather high even though maximum NPP is reached quite early probably because of the high stand density and the absence of thinnings. Magnani et al. (2005) reported an aboveground NPP of 1.8 Mg C ha^{-1} y^{-1} for a 12 year old plantation. The same authors showed that maximum NEE was reached at 9–10 years. In our plantation, at the same age aboveground NPP was 2.7 Mg C ha^{-1} y^{-1}. Considering all the realized plantations (1992–2006), mean annual aboveground NPP is 2.4 Mg C ha^{-1} y^{-1}, a value lower than 4.3 Mg C ha^{-1} y^{-1} estimated by Anderle et al. (2002).

The average density per hectare in poplar plantations was around 204 stems ha^{-1}. Aboveground carbon stock ranged between 2.7 and 56.1 Mg C ha^{-1} in 3 and 9 year old stands, respectively. A good agreement between measured values and values reported by the yield table for Veneto (Famiglietti 1968) was found (Fig. 13.2b). Aboveground NPP, derived from the yield table, was between 0.9 and 12.0 Mg C ha^{-1} y^{-1}, underlying the very high productivity of this species.

A good correlation between soil carbon and aboveground carbon stocks was found in both chronosequences (Table 13.2). Soil carbon stock ranged between 49.8 and 56.9 Mg C ha^{-1} at 0 and 30 years in mixed broadleaf plantations and

Table 13.1 Average carbon stock and average annual stock change in different pools (aboveground, roots and soil) for mixed broadleaf and poplar plantations in Friuli Venezia Giulia

		Aboveground	Roots	Soil
Carbon stock (Mg C ha^{-1})	Mixed broadleaf	25.3	6.6	54.3
	Poplar	40.3	10.5	77.0
Carbon flux (Mg C ha^{-1} y^{-1})	Mixed broadleaf	2.4	0.6	0.4
	Poplar	9.4	2.4	1.1

Table 13.2 Regression coefficients between soil carbon stock and aboveground carbon stock in both the chronosequences (Y = a X + b)

	a	b	R^2	n	*p*-value
Mixed broadleaf	0.18	49.73	0.96	4	0.02
Poplar	1.12	68.58	0.76	5	0.06

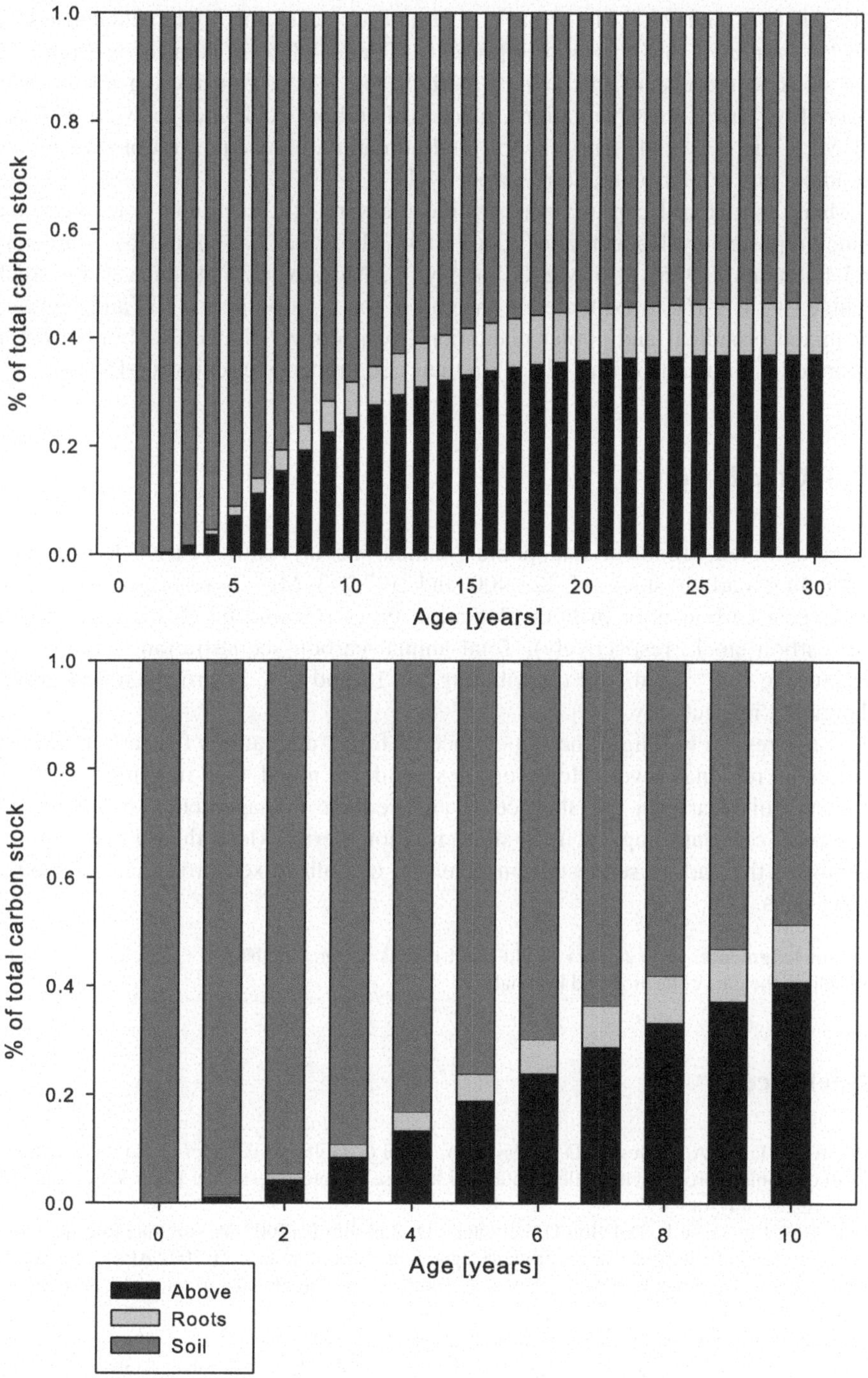

Fig. 13.3 Total carbon stock partitioning among different ecosystem pools with stand age. *Top panel* broadleaf plantations; *bottom panel* poplar plantation

between 68.6 and 79.8 Mg C ha^{-1} at 0 and 10 years for poplar plantations. It can be estimated that this pool represents 53 and 48 % of total ecosystem carbon stock at the end of the rotation period (Fig. 13.3). Considering all the area covered by such forest plantations, mean soil carbon stock changes were 0.4 and 1.1 Mg C ha^{-1} y^{-1} (13 and 9 % of total annual carbon sequestration) for mixed broadleaf and poplar plantations, respectively.

Mean annual soil carbon sequestration measured in this study is between the values reported by Rodeghiero (2006) (0.33 Mg C ha^{-1} y^{-1}) and by Bouwman and Leemans (1995) (1.6 Mg C ha^{-1} y^{-1}). Moreover, final soil carbon stock change (+14, +16 %) estimated at the end of the rotation period (30 and 10 years for mixed broadleaf and poplar plantations, respectively) is well within the range reported by Guo and Gifford (2002) in their comprehensive review (+18 %).

13.4 Conclusions

Overall, mixed broadleaf and poplar plantations in Friuli Venezia Giulia have a standing carbon stock of 223,400 and 397,753 Mg C, respectively. Soil is the largest carbon pool in both plantation types (60 and 63 % of total ecosystem carbon stock, respectively). Total annual carbon sequestration is 8,855 and 40,286 Mg C y^{-1} with soil contributing for 13 and 9 % in broadleaf and poplar plantation, respectively.

These results highlight the importance of forest plantations for carbon sequestration at regional level. However, it should be noted that maximum NPP is reached quite early in the absence of an adequate management (i.e. thinnings). However, our data suggest that short rotation period (less than 12 years) can maximize the carbon sequestration potential of both mixed broadleaf and poplar plantations.

Acknowledgments The authors would like to thank Diego Chiabà and Gabriele Pingitore for the help in the data collection and lab analysis.

References

Alberti G, Marelli A, Piovesana D, Peressotti A, Zerbi G, Gottardo E, Bidese F (2006) Accumulo di carbonio e produttività delle piantagioni legnose (Kyoto forests) del Friuli Venezia Giulia. Forest@ 3(4):488–495

Anderle A, Ciccarese L, Dal Bon D, Pettenella D, Zanolini E (2002) Assorbimento e fissazione del carbonio nelle foreste e nei prodotti legnosi in Italia. Rapporto 21/2002 APAT, Roma p 58

Bouwman AF, Leemans R (1995) The role of forest soil in the global carbon cycle. Soil Sci Soc Am 503–525

Cooper CF (1983) Carbon storage in managed forests. Can J For Res 13:155–166

Elliott ET, Heil JW, Kelly EF, Monger HC (1999) Soil structural and other physical properties. In: Robertson GP, Coleman DC, Bledsoe CS, Sollins P (eds) Standard soil methods for long-term ecological research. Oxford University Press, Oxford, pp 74–85

Famiglietti A (1968) Tavola alsometrica del pioppo per il Veneto. In: Castellani C (ed) Tavole stereometriche ed alsometriche costruite per i boschi Italiani. Istituto Sperimentale per l'Assestamento Forestale e per l'Alpicoltura: 693

Gasparini P, Tabacchi G. (a cura di) (2011) L'Inventario Nazionale delle Foreste e dei serbatoi forestali di Carbonio INFC 2005. Secondo inventario forestale nazionale italiano. Metodi e risultati. Ministero delle Politiche Agricole, Alimentari e Forestali; Corpo Forestale dello Stato. Consiglio per la Ricerca e la Sperimentazione in Agricoltura, Unità di ricerca per il Monitoraggio e la Pianificazione Forestale. Edagricole-Il Sole 24 ore, Bologna. pp 653

Guo LB, Gifford RM (2002) Soil carbon stocks and land use change: a meta analysis. Glob Change Biol 8(4):345–360

Hooker TD, Compton JE (2003) Forest ecosystem carbon and nitrogen accumulation during the first century after agricultural abandonment. Ecol Appl 13(2):299–313

Houghton RA (2003) Why are estimates of the terrestrial carbon balance so different. Glob Change Biol 9(4):500–509

Houghton RA, Hackler JL (2000) Changes in terrestrial carbon storage in the United States. I: the roles of agriculture and forestry. Glob Ecol Biogeogr 9(2):125–144

IPCC (2000) In: Watson RT, Noble IR, Bolin B, Ravindranath NH, Verardo DJ, Dokken DJ (eds) Land use, land-use change and forestry. Cambridge University Press, Cambridge, p 375

IPCC (2003) Penman J, Gytarsky M, Hiraishi T, Krug T, Kruger D, Pipatti R, Buendia L, Miwa K, Ngara T, Tanabe K, Wagner F, Good Practice Guidance for Land Use, land-Use Change and Forestry IPCC/IGES, Hayama, Japan

Magnani F, Grassi G, Tonon G, Cantoni L, Ponti F, Vicinelli E, Boldreghini P, Nardino M, Georgiadis T, Facini O, Rossi F (2005) Quale ruolo per l'arboricoltura da legno italiana nel protocollo di Kyoto? Indicazioni da una "Kyoto foret" della pianura emiliana. Forest@ 2(4):333–344

Pacala SW, Hurtt GC, Baker D, Peylin P, Houghton RA, Birdsey RA, Heath L, Sundquist ET, Stallard RF, Ciais P, Moorcroft P, Caspersen JP, Shevliakova E, Moore B, Kohlmaier G, Holland E, Gloor M, Harmon ME, Fan SM, Sarmiento JL, Goodale CL, Schimel D, Field CB (2001) Consistent land-and atmosphere-based US carbon sink estimates. Science 292(5525):2316–2320

Paul KI, Polglase PJ, Nyakuengama JG, Khanna PK (2002) Change in soil carbon following afforestation. For Ecol Manag 168:241–257

Rodeghiero M (2006) Metodologie per la stima del carbonio nei suoli forestali. In: Pilli R, Anfodillo T, Dalla Valle E (eds) Stima del Carbonio in foresta: metodologie ed aspetti normativi, Pubblicazione del Corso di Cultura in Ecologia, Atti del 42°corso, Università di Padova, pp 141–159

Chapter 14
Trying to Link Vegetation Units with Biomass Data: The Case Study of Italian Shrublands

Salvatore Pasta, Tommaso La Mantia, Serena Marras, Costantino Sirca, Donatella Spano and Riccardo Valentini

Abstract Although their carbon stock is relevant in assessing the baseline for the negotiation of future agreements with respect to carbon balance, there still are few available studies concerning the biomass and the net ecosystem exchange capacity of Mediterranean shrublands. In this chapter a preliminary overview on the biomass values concerning Italian shrubland communities and/or their dominant/characteristic woody species is provided. Many useful data on above- and below-ground biomass issued from investigations carried out in other Mediterranean countries and concerning plant communities, which share the same ecological, floristic and structural traits of Italian shrublands. A preliminary finding of this research is the uneven degree of knowledge concerning the different non-forest woody communities. For example, there is still no literature on the biomass of some 2/3 of all the considered phytosociological units. Besides, both the above and the below-ground biomass of many Mediterranean shrubs show a very wide range of variation as they are strongly influenced by progressive succession processes and by the nature, the intensity and the frequency of disturbance factors. Thus, direct measuring of these values for each vegetation unit and dominant woody species should be encouraged and intensified. Monitoring activities concerning biomass increase are recommended as well: as a matter of fact, at present

S. Pasta · T. La Mantia (✉)
Department of Agricultural and Forestry Sciences, University of Palermo, Palermo, Italy

S. Marras · C. Sirca · D. Spano
Department of Natural and Territorial Science, University of Sassari, Sassari, Italy

S. Marras · C. Sirca · D. Spano
Impacts on Agriculture, Forest and Natural Ecosystem Division (IAFENT), Euro-Mediterranean Center on Climate Changes (CMCC), Sassari, Italy

R. Valentini
Impacts on Agriculture, Forest and Natural Ecosystem Division (IAFENT), Euro-Mediterranean Center on Climate Changes (CMCC), Viterbo, Italy

R. Valentini and F. Miglietta (eds.), *The Greenhouse Gas Balance of Italy*, Environmental Science and Engineering, DOI 10.1007/978-3-642-32424-6_14

reference data on this topic are so limited and variable that it is not possible to confidently estimate the annual growth of shrubland communities.

14.1 Introduction

Shrublands may represent the most mature local communities and not be able to "evolve" towards proper forests. In these cases, their presence is mostly due to very strong limiting (stress) factors, such as long-lasting warm or cold periods, scarce water availability, nutrient-poor or toxic soils, intensive drainage, etc. Elsewhere, shrublands represent unsteady steps within a dynamic process, which may be progressive or retrogressive. In the first case, they issue from the recent colonization performed by woody species after land abandonment (Corona et al. 2008; La Mantia et al. 2008), and their complexity increases along with cover rate till they become mature forests. In retrogressive succession, in contrast, severe and/or frequent disturbance (i.e., grazing, burning, cutting, etc.) causes the structural simplification of pre-existing forest communities. As a consequence, shrublands and open maquis communities may have similar physiognomies even when they have rather different histories to tell. Thus, a lack of historical information may bias any trial to assess how plant biomass is likely to change in the future. On the other hand, as shrub communities often develop when forests degrade, the knowledge of their carbon stock is relevant in assessing the baseline for negotiating future agreements with respect to carbon balance.

At present the official data concerning non-forest Italian vegetation do not match perfectly: in fact, according to the most recent (2005) National Forestry Inventory (http://www.sian.it/inventarioforestale/jsp/home.jsp) shrublands cover 990,916 ha and woodlands 46,678 ha, while following the official data provided by the National Forestry Inventory carried out in 1985, Corona et al. (1997) estimate that altogether Italian shrubby, rupicolous, and riparian plant communities occupy 2,164,500 ha and have a carbon stock of 26.4 (18–35) Mt. The same authors also report 256 Mt as the average value of carbon stock in Italian forests (high forests + coppices) and tree plantations.

Beyond these discrepancies, there is no doubt about the important role played by shrublands in the global carbon stock, as indicated by the Kyoto Protocol (the "Marrakesh Agreements" explicitly refer to shrub vegetation colonizing abandoned agricultural land). On the other hand, there are few available studies concerning the biomass and the net ecosystem exchange capacity of shrublands (Navarro Cerrillo and Blanco Oyonarte 2006). This is also true for the most common woody species (hereinafter abbreviated "WS") typical to Mediterranean (Usò et al. 1997) and Italian (Costa and La Mantia 2005; Corona et al. 2012) shrublands, and even less information is available concerning biomass assessment at the community level. This knowledge gap not only depends on the lack of field investigations focused on the dominant WS but also on the high floristic and structural variability of shrublands themselves.

Besides, most of Mediterranean WS show high size variability (Corona et al. 2012 and references within), so that the common techniques and algorithms used for the forest tree species do not always fit well with maquis ones. Even ring analysis techniques often fail for these species because high environmental variability makes interpretation of ring data difficult (Cherubini et al. 2003). Because a standardised methodology is unavailable for these species, researchers have estimated maquis biomass and the annual variation in the biomass using a variety of methods. As a result, the available data are seldom comparable (Corona et al. 2012).

During the last four decades, plant ecologists have published hundreds of papers and thousands of phytosociological relevés concerning the non-forest Italian plant communities. This huge amount of data is mostly focused on floristic similarities among coenoses, but many of the papers provide only basic information on the abiotic context (e.g., soil chemistry, bioclimate, slope, aspect, disturbance intensity and frequency, etc.), and very few report the main structural patterns (vertical stratification, cover of each vegetation layer, etc.). This makes difficult to compare the relevés published in different works, even if they concern the same plant community. In this chapter an attempt to interpret phytosociological data has been carried out in order to provide a preliminary overview on the biomass values of Italian shrubland communities and/or their dominant/characteristic WS.

14.2 Basic Assumptions on Italian Shrubland Vegetation

Several thousands of phytosociological relevés concerning the Italian non-forest woody communities published till June 2011 were considered (data not shown). In order to facilitate modelling, we focused our attention on the similarities concerning the following six parameters: (1) whole vascular flora (all taxa); (2) 'weighted' vascular flora (all taxa also considering coverage values); (3) dominant WS; (4) most frequent WS (i.e. whose frequence was $\geq$ 50 %); (5) ecological (i.e., species with similar edapho-climatic requirements); and (6) structural (i.e. cover rate, stratification, etc.) patterns.

The nomenclature of vegetation units follows Biondi and Blasi (2009). All those classes in which WS are dominant or frequent in only one or a few associations, like the pioneer vegetation of screes and ravines (class *Thlaspietea rotundifolii*), the chasmophilous communities of undisturbed cliffs (class *Asplenietea trichomanis*), the meso-xerophilous perennial grasslands (class *Festuco-Brometea*), the chamaephytic lithophilous communities prone to salt spray (class *Crithmo-Staticetea pro parte*), the rush-dominated prairies on salt-rich and humid soils (class *Juncetea maritimi*) were excluded. The same decision was taken for high maquis formations dominated by *Quercus ilex*, *Quercus suber*, and *Pinus* spp. because they should be ascribed to proper woodland communities (class *Quercetea ilicis*, order *Quercetalia ilicis*).

Many useful data on above- and belowground biomass (hereinafter Ab and Bb) issued from investigations which have been carried out in other Mediterranean countries (e.g. Spain, southern France, Greece and Israel) and concerned plant communities which shared the same ecological, floristic and structural traits of Italian shrublands.

14.3 The Study Case of Italian Shrublands

Italian shrubland plant communities are ascribed to 17 classes, 26 orders, 56 alliances (Table 14.1) and c. 730 associations and subassociations (data not shown). The most widespread communities resulted to be: (1) the Mediterranean maquis and scrubland (class *Quercetea ilicis*, order *Quercetalia calliprini*); (2) the submediterranean deciduous mantle assemblages (class *Rhamno cathartici-Prunetea spinosae*); (3) the acidophilous (classes *Cisto ladaniferi-Lavanduletea stoechadis* and *Cytisetea scopario-striati*) and basiphilous (*Cisto cretici-Micromerietea julianae* and *Rosmarinetea officinalis*) garrigues and heaths; (4) the coastal sub-halophilous shrublands and garrigues (*Crithmo-Staticetea pro parte*); (5) the psammophilous garrigues typical to the fixed dunes (*Helichryso italici-Crucianelletea maritimae pro parte*); (6) the alpine, sub-alpine, and apennine scrublands and heaths (classes *Erico-Pinetea pro parte*, *Calluno-Ulicetea* and *Pino-Juniperetea*); (7) the oromediterranean heath communities typical to the mountain tops of Central and Southern Italy, Sardinia, and Sicily (*Carici-Genistetea lobelii* and *Rumici-Astragaletea siculi*); (8) the thermo-hygrophilous open riparian communities (class *Nerio-Tamaricetea*) and (9) the halo-xerophilous (class *Pegano-Salsoletea*) and halo-xeronitrophilous (class *Sarcocornietea fruticosae*) scrub communities dominated by shrubby chenopods.

Table 14.2 provides a synthesis of the useful data on above- and belowground biomass concerning (or which could be applied to) the Italian shrubland communities and their dominant/characteristic WS.

A preliminary finding of our research is the uneven degree of knowledge concerning the different shrubland Italian plant communities. For example, many papers concern the thermo-xerophilous maquis assemblages referred to a single phytosociological alliance, i.e. *Oleo-Ceratonion*, while there is still no literature on the biomass of 11 out of 17 classes, 18 out of 26 orders, and 39 out of 57 alliances.

More in detail, no information was found concerning *Nerio-Tamaricetea*, *Cytisetea scopario-striati*, *Calluno vulgaris-Ulicetea minoris*, *Erico-Pinetea*, *Vaccinio-Piceetea* and *Pino-Juniperetea* and most part of the WS-dominated plant communities typical to coastal habitats and referred to the classes *Crithmo-Staticetea*, *Helichryso-Crucianelletea maritimae*, *Pegano-Salsoletea* and to the order *Limonietalia* of the class *Sarcocornietea fruticosae*.

Many of the plant communities whose biomass has never been evaluated play only a minor role in terms of biomass because they cover rather small surfaces and/or because they are dominated by small shrubs (e.g. *Rumici-Astragaletea siculi* and *Carici-Genistetea lobelii*). The only important exceptions are the mantle communities referred to *Rhamno cathartici-Prunetea spinosae*: even though they

Table 14.1 Overview of the dominant and/or most frequent WS within an up-to-date syntaxonomic scheme of Italian pre-forest vegetation

Phytosociological unit			Dominant WS	Other characteristic WS	Importance–biomass (see note 1)	Available data on biomass estimation (cfr. Table 14.2)
Class	Order	Alliance				
Crithmo-Staticetea	*Crithmo-Limonietalia* (incl. *Senecionetalia cinerariae*)	*Plantagini-Thymelaeion hirsutae*	*Thymelaea hirsuta*	*Helichrysum* spp., *Senecio cineraria s.l.*	M–M	N
		Anthyllidion barbae-jovis	*Anthyllis barba-jovis*	*Helichrysum* spp.	M–M	N
		Crucianellion rupestris	*Crucianella rupestris*	*Limonium* spp.	L–L	N
Sarcocornietea fruticosae	*Sarcocornietalia fruticosae*	*Sarcocornion alpinii*	*Sarcocornia alpinii*		M–M	N
		Sarcocornion fruticosae	*Sarcocornia fruticosa*	*Halimione portulacoides*	M–M	Y
		Arthrocnemion macrostachyi	*Arthrocnemum macrostachyum*	*Halocnemum macrostachyum*	M–L	Y
		Suaedion verae	*Suaeda vera*	*Halimione portulacoides*, *Limoniastrum monopetalum*	M–M	Y
	Limonietalia	*Limoniastrion monopetali* (= *Limonion ferulacei*)	*Limonium ferulaceum*	*Limoniastrum monopetalum*	L–L	N

(continued)

Table 14.1 (continued)

Phytosociological unit			Dominant WS	Other characteristic WS	Importance–biomass (see note 1)	Available data on biomass estimation (cfr. Table 14.2)
Class	Order	Alliance				
Cisto ladaniferi-Lavanduletea stoechadis	*Lavanduletalia stoechadis*	*Cistion ladaniferi*	*Lavandula stoechas*	*Cistus* spp., *Erica scoparia*	M–L	Y
		Calicotomo-Genistion tyrrhenae	*Calicotome* spp., *Genista* gr. *ephedroides*	*Cytisus villosus, Genista monspessulana*	M–M	N
		Teucrion mari	*Teucrium marum, Genista* spp.	*Cistus* spp., *Rosmarinus officinalis, Halimium halimifolium*	M	N
Helichryso-Crucianelletea maritimae	*Helichryso-Crucianelletalia maritimae*	*Crucianellion maritimae*	*Crucianella maritima*	*Ephedra distachya, Helichrysum microphyllum, Thymelaea tartonraira*	L–L	N
		Euphorbion pithyusae	*Euphorbia pithyusa* subsp. *pithyusa*	*Helichrysum microphyllum, Astragalus thermensis*	L–L	N
Cisto cretici-Micromerietea julianae	*Cisto cretici-Ericetalia manipuliflorae*	*Cisto cretici-Ericion manipuliflorae*	*Cistus creticus, Erica manipuliflora*	*Sarcopoterium spinosum*	M–M	Y
Rosmarinetea officinalis	*Ononidetalia striatae*	*Lavandulo-Genistion cinereae*		*Lavandula stoechas, Genista cinerea*	M–L	Y
	Rosmarinetalia officinalis	*Rosmarinion officinalis*	*Rosmarinus officinalis*	*Globularia alypum*	M–M	Y
		Aphyllanthion	*Aphyllanthes monspeliensis*		L–L	N
		Cisto erio-cephali-Ericion multiflorae	*Cistus eriocephalus, Erica multiflora*	*Sarcopoterium spinosum*	M–M	Y
		Euphorbion ligusticae	*Euphorbia ligustica*		L–L	N

(continued)

Table 14.1 (continued)

Phytosociological unit			Dominant WS	Other characteristic WS	Importance–biomass (see note 1)	Available data on biomass estimation (cfr. Table 14.2)
Class	Order	Alliance				
Cytisetea scopario-striati	*Cytisetalia scopario-striati*	*Violion messanensis*	*Sarothamnus scoparius*	*Adenocarpus* spp.	M–L	N
	Cytiso villosi-Telinetalia monspessulanae	*Telinion monspessulano-linifoliae*	*Genista monspessulana, Cytisus villosus*	*Ulex europaeus*	M–L	N
Rhamno cathartici-Prunetea spinosae	*Prunetalia spinosae*	*Cytision sessilifolii*	*Cytisus sessilifolius*	*Crataegus laevigata, Amelanchier ovalis*	M–M	N
		Pruno-Rubion ulmifolii	*Rubus ulmifolius, Prunus spinosa*	*Crataegus* spp., *Rosa* spp., *Tamus communis, Smilax aspera, Lonicera etrusca, Rhus coriaria, Amelanchier ovalis, Cotinus coggygria, Juniperus oxycedrus, Juniperus communis, Sambucus nigra, Pyrus* spp., *Paliurus spina-christi, Euphorbia characias, Sambucus nigra, Prunus mahaleb, Cornus sanguinea, Coronilla emerus, Pistacia terebinthus, Ulmus minor, Corylus avellana*	H–H	N
		Sarothamnion scoparii	*Sarothamnus scoparius*		M–L	N
		Cytiso spinescentis-Saturejion montanae	*Chamaecytisus spinescens, Satureja montana*		L–L	N
		Berberidion vulgaris	*Berberis vulgaris*	*Ligustrum vulgare, Sambucus nigra, Amelanchier ovalis, Cotinus coggygria, Juniperus oxycedrus, Paliurus spina-christi, Cornus* spp., *Prunus mahaleb, Frangula* spp., *Rhamnus* spp.	M–H	N

(continued)

Table 14.1 (continued)

Phytosociological unit			Dominant WS	Other characteristic WS	Importance–biomass (see note 1)	Available data on biomass estimation (cfr. Table 14.2)
Class	Order	Alliance				
Quercetea ilicis	*Quercetalia calliprini*	*Oleo-Ceratonion siliquae*	*Olea europaea* var. *sylvestris*, *Pistacia lentiscus*	*Phillyrea* spp., *Myrtus communis*, *Euphorbia dendroides*, *Chamaerops humilis*, *Clematis cirrhosa*, *Rhamnus alaternus*, *Anagyris foetida*, *Ephedra fragilis*, *Quercus calliprinos*, *Teucrium fruticans*, *Bupleurum fruticosum*, *Ephedra fragilis*, *Zizyphus lotus*, *Rhamnus oleoides*, *Phlomis fruticosa*	H–H	Y
		Ericion arboreae	*Erica arborea*	*Arbutus unedo*, *Cistus* spp.	H–M	Y
		Periplocion angustifoliae	*Periploca angustifolia*	*Euphorbia dendroides*, *Juniperus turbinata*, *Rhus tripartita*	M–L	N
		Juniperion turbinatae	*Juniperus turbinata*, *Juniperus macrocarpa*	*Phillyrea angustifolia*, *Pistacia lentiscus*, *Chamaerops humilis*, *Quercus calliprinos*, *Ephedra fragilis*, *Olea europaea* var. *sylvestris*, *Juniperus communis* subsp. *communis*	H–M	Y
Nerio-Tamaricetea	*Tamaricetalia*	*Tamaricion africanae*	*Tamarix* spp.		M–M	N
		Rubo ulmifolii-Nerion oleandri	*Rubus ulmifolius*, *Nerium oleander*	*Vitex agnus-castus*, *Spartium junceum*	M–M	N

(continued)

Table 14.1 (continued)

Phytosociological unit			Dominant WS	Other characteristic WS	Importance–biomass (see note 1)	Available data on biomass estimation (cfr. Table 14.2)
Class	Order	Alliance				
Pino-Juniperetea	*Pino-Juniperetalia*	*Daphno oleoidis-Juniperion alpinae*	*Juniperus alpina*		M–M	N
	Juniperetalia hemisphaericae	*Berberidion aetnensis*	*Berberis aetnensis*	*Juniperus nana*	M–M	N
		Juniperion nanae	*Juniperus nana*		M–M	N
		Juniperion thuriferae	*Juniperus thurifera*	*Juniperus communis, J. hemisphaerica*	M–L	N
		Epipactido atropurpureae-Pinion mugo	*Pinus mugo*	*Polygala chamaebuxus*	M–L	N
Calluno vulgaris-Ulicetea minoris	*Vaccinio myrtilli-Genistetalia pilosae*	*Genistion pilosae*	*Genista pilosa*	*Chamaecytisus hirsutus, Calluna vulgaris*	L–L	N
	Ulicetalia minoris	*Cisto salvifolii-Ericion cinereae*	*Cistus salvifolius, Erica cinerea*	*Calluna vulgaris*	L–L	N

(continued)

Table 14.1 (continued)

Phytosociological unit			Dominant WS	Other characteristic WS	Importance–biomass (see note 1)	Available data on biomass estimation (cfr. Table 14.2)
Class	Order	Alliance				
Rumici-Astragaletea siculi	*Rumici aetnensis-Astragaletalia siculi*	*Rumici aetnensis-Astragalion siculi*	*Astragalus siculus*		L–L	N
	Erysimo bonanniani-Jurinetalia bocconei	*Cerastio tomentosi-Astragalion nebrodensis*	*Astragalus nebrodensis*	*Thymus spinulosus, Helichrysum italicum*	L–L	N
		Armerion nebrodensis	*Armeria nebrodensis*	*Genista cupanii*	L–L	N
	Anthemidetalia calabricae	*Koelerio-Astragalion calabrici*	*Astragalus calabricus*	*Genista silana, Chamecytisus spinescens*	L–L	N
		Armerion aspromontanae	*Armeria aspromontana*		L–L	N
Carici-Genistetea lobelia	*Teucrio-Santolinetalia insularis*	*Anthyllidion hermanniae*	*Anthyllis hermanniae*	*Genista salzmannii, Helichrysum microphyllum, Astragalus genargenteus, Genista corsica, Genista lobelioides*	L–L	N
		Armerio sardoae-Genistion salzmannii	*Armeria sardoa, Genista salzmannii*	*Santolina insularis*	L–L	N
		Polygalo-Seslerion insularis		*Helianthemum croceum, Santolina insularis*	L–L	N

(continued)

Table 14.1 (continued)

Phytosociological unit			Dominant WS	Other characteristic WS	Importance–biomass (see note 1)	Available data on biomass estimation (cfr. Table 14.2)
Class	Order	Alliance				
Erico-Pinetea	*Erico-Pinetalia*	*Ericion carneae*	*Erica carnea*		L–L	N
		Erico-Pinion mugo	*Erica carnea, Pinus mugo*	*Rhododendrum hirsutum, Arctostaphylos uva-ursi, Sorbus chamaemespilus*	M–L	N
		Erico-Fraxinion orni	*Erica carnea, Fraxinus ornus*	*Amelanchier ovalis*	M–L	N
Vaccinio-Piceetea	*Vaccinio-Piceetalia*	*Vaccinio-Piceion*	*Vaccinium* spp.		M–L	N
	Rhododendro-Vaccinietalia	*Rhododendro-Vaccinion*	*Rhododendrum ferrugineum, R. hirsutum, Vaccinium gaultheriodies, V. vitis-idaea, V. myrtillus*	*Calluna vulgaris*	L–L	N
		Loiseleurio-Vaccinion	*Vaccinium* spp.		L–L	N
Pegano-Salsoletea	*Salsolo vermiculatae-Peganetalia harmalae*	*Salsolo vermiculatae-Peganion harmalae*	*Salsola vermiculata s.l., Salsola oppositifolia*	*Suaeda vera, Halimione portulacoides, Atriplex halimus, Limonium* spp.	M–M	N
		Artemision arborescentis	*Artemisia arborescens*	*Suaeda vera, Atriplex halimus*	M–M	N
	Helichryso-Santolinetalia	*Artemision variabilis*	*Helichrysum italicum, Santolina* spp.	*Senecio cineraria s.l.*	L–L	N

Knowledge level: H = high; M = medium; L = low

Note: to evaluate the relative importance of each considered alliance either the distribution and/or the aboveground biomass (Ab) were taken into account: H (=high): very widespread plant community and/or with high Ab; M (=medium); L (=low)

Table 14.2 Data on the aboveground (Ab) and the belowground (Bb) biomass values (g/m^2; ° = g/individual) of Italian shrubland plant communities and WS obtained from investigations carried out within the national territory or obtained through comparisons with similar European study cases

Considered phytosociological units	Considered dominant-characteristic WS
Sarcocornietea fruticosae Ab: 952 (Curcó et al. 2002); Ab: 302 (García et al. 1993)	*Sarcocornia fruticosa* Ab: 678 (Scarton et al. 2002); 1,297 (Scarton 2006); 588–1,198–3,448 (Curcò et al. 2002); 581–1,297–1,800 (Scarton 2006); 598 (Laffoley and Grimsditch 2009); Bb: 2,829 (Scarton et al. 2002); 1,601 (Laffoley and Grimsditch 2009) *Sarcocornia perennis* Ab: 450–550–600 (Curcò et al. 2002) *Arthrocnemum macrostachyum* Ab: 311–400–872 (Curcò et al. 2002); 190–683–840 (Laffoley and Grimsditch 2009) 928 ± 79 (Neves et al. 2010); Bb: 50–340–1,260 (Laffoley and Grimsditch 2009); 1,070 ± 50 (Neves et al. 2010) *Halimione portulacoides* Ab: 561–1,270–1,541 (Scarton 2006); 598 (Neves et al. 2007); Bb: 1,601 (Neves et al. 2007)
Cisto cretici-Micromerietea julianae Bb: 1,618 (Margaris 1976)	*Phlomis fruticosa* Ab: 366.5 (Margaris 1976) *Sarcopoterium spinosum* Ab: 143.5 (Margaris 1976) *Coridothymus capitatus* Ab: 36.0 (Margaris 1976); 226.12 ± 313.8° (La Mantia unpubl.); 10.0 (Bianchi et al. 2002); Bb: 69.34 ± 84.6° (La Mantia unpubl.) *Cistus* spp. Ab: 83.5 (Margaris 1976) *Helianthemum* spp. Ab: 80.3 (Margaris 1976) *Asparagus aphyllus* Ab: 11.6 (Margaris 1976) *Micromeria fruticulosa* Ab: 200.53 ± 199° (La Mantia unpubl.); Bb: 96.12 ± 74.0° (La Mantia unpubl.)
Rosmarinetea officinalis Ab: 2,142–3,106 and <2,500 (Navarro Cerrillo and Blanco Oyonarte 2006); 433–827 (Calvo 2007)	*Rosmarinus officinalis* Ab: 111.4° (Usò et al. 1997); 46.5–1,828° (Blanco Oyonarte and Navarro Cerrillo 2003) *Genista cinerea*: 307.9–6,745° (Blanco Oyonarte and Navarro Cerrillo 2003) *Helichrysum stoechas* Ab: 128.7° (Usò et al. 1997) *Cistus albidus* Ab: 100.5° (Usò et al. 1997); 74.6–762° (Blanco Oyonarte and Navarro Cerrillo 2003) *Teline linifolia* Ab: 64.4–923.4° (Blanco Oyonarte and Navarro Cerrillo 2003)
Cisto ladaniferi-Lavanduletea stoechadis Ab: 447 ± 135; 2,482 (Navarro Cerrillo and Blanco Oyonarte 2006); 194.71–1,164.89 (Castro and Freitas 2009)	*Cistus ladanifer*: 788 ± 232; 3,030 (Navarro Cerrillo and Blanco Oyonarte 2006) *Helichrysum stoechas*: 30.0 (Bianchi et al. 2002) *Lavandula stoechas*: 49.5–636.8° (Blanco Oyonarte and Navarro Cerrillo 2003) *Cistus monspeliensis*: 10.0–80.0 (Bianchi et al. 2002); 177.5–2,776° (Blanco Oyonarte and Navarro Cerrillo 2003) *Cistus salvifolius*: 50.0–80.0 (Bianchi et al. 2002) *Genista corsica*: 90–130 (Bianchi et al. 2002) *Halimium halimifolium*: 30 (Bianchi et al. 2002)
Calluno vulgaris-Ulicetea minoris Ab 1,143 ± 736; 1,404 ± 215 (Navarro Cerrillo and Blanco Oyonarte 2006)	

(continued)

Table 14.2 (continued)

Considered phytosociological units	Considered dominant-characteristic WS
Rhamno cathartici-Prunetea spinosae	*Spartium junceum*: 160–320 (Corona et al. 2012) *Crataegus monogyna* Ab: 112.4–7,419° (Blanco Oyonarte and Navarro Cerrillo 2003)
Quercetea ilicis Quercetalia calliprini Ab: 1,010 (Peressotti et al. 1999); 695–248 (Navarro Cerrillo and Blanco Oyonarte 2006); 1,700 (Sağlam et al. 2008) *Oleo-Ceratonion siliquae* Ab: 8,387 (Gratani et al. 1980); 3,380 (Catarino et al. 1982); 1,496 (Hilbert and Canadell 1995); 1,514–1,689 (Navarro Cerrillo and Blanco Oyonarte 2006); Bb: 7,200 (Hilbert and Canadell 1995) *Ericion arboreae* Ab: 2000–2500 (Bianchi et al. 2002); 1,000–6,000 (Navarro Cerrillo and Blanco Oyonarte 2006) *Juniperion turbinatae* Ab: 4,425 (Gratani et al. 1980) *Periplocion angustifoliae*	*Quercus coccifera* Ab: 2,350 (Rapp and Loissant 1981); 1,000–5,000 (Navarro Cerrillo and Blanco Oyonarte 2006); Bb: 4,600 (Rapp and Loissant 1981) *Olea europaea* var. *sylvestris* Ab: 1,000 (Navarro Cerrillo and Blanco Oyonarte 2006) *Pistacia lentiscus* Ab: 1,966 ± 617 (Navarro Cerrillo and Blanco Oyonarte 2006); 140–280 (Corona et al. 2012); 130 (Peressotti et al. 1999) *Euphorbia dendroides* Ab: 130–260 (Corona et al. 2012) *Chamaerops humilis* Ab: 140–280 (Corona et al. 2012); 224.2–5,735° (Blanco Oyonarte and Navarro Cerrillo 2003) *Calicotome villosa* Ab: 1,786–17,940° (Blanco Oyonarte and Navarro Cerrillo 2003) *Daphne gnidium* Ab: 42–1,258° (Blanco Oyonarte and Navarro Cerrillo 2003) *Juniperus oxycedrus* Ab: 117.9–5,392° (Blanco Oyonarte and Navarro Cerrillo 2003) *Myrtus communis* Ab: 203–5,380° (Blanco Oyonarte and Navarro Cerrillo 2003) *Pistacia terebinthus* Ab: 1,666–24,861° (Blanco Oyonarte and Navarro Cerrillo 2003) *Rhamnus oleoides* Ab: 30–9,278° (Blanco Oyonarte and Navarro Cerrillo 2003) *Teucrium fruticans* Ab: 4.4–81.5° (Blanco Oyonarte and Navarro Cerrillo 2003) *Viburnum tinus* Ab: 105.1–2,617.2° (Blanco Oyonarte and Navarro Cerrillo 2003) *Erica arborea* Ab: 870–2580 (Bianchi et al. 2002); 6.3–1,860° (Blanco Oyonarte and Navarro Cerrillo 2003) *Arbutus unedo* Ab: 1,050 (Bianchi et al. 2002); 15.4–35,760° (Blanco Oyonarte and Navarro Cerrillo 2003) *Phillyrea angustifolia* Ab: 0.1 (Bianchi et al. 2002); 45.2–1,920° (Blanco Oyonarte and Navarro Cerrillo 2003) *Juniperus phoenicea* Ab: 790 (Peressotti et al. 1999); 109.7–8,153° (Blanco Oyonarte and Navarro Cerrillo 2003) *Lycium intricatum* Ab: 1,200 (Otto et al. 2001) *Periploca laevigata* Ab: 900 (Otto et al. 2001)
Pino-Juniperetea	*Juniperus thurifera* Ab: 2,700–5,800–6,700–20,900 (average value 4,900) (Montès et al. 2002)

Note: the data concerning the different alliances of the order *Quercetalia calliprini* (i.e. *Oleo-Ceratonion*, *Ericion arboreae*, *Juniperion turbinatae* and *Periplocion angustifoliae*) are presented separately due to the high variability of their floristic and structural patterns occupy discontinuous areas, their biomass is very significant at national level. Nevertheless, the few available data concerning some of the most common mantle-dominating species can be used to assess community biomass through modeling.

14.4 Discussion and Research Perspectives

All Italian shrubland plant communities may be consolidated within few 'landscape macrocategories' through multivariate statistics, a task that is still in progress. Thus, we shall avow that at present it is still hard to obtain inequivocal proxies to predict the productivity of all Italian shrubland ecosystems. Nonentheless, available information arising from field investigations already carried out abroad allows to estimate the range of variation of the biomass values of some phytosociological classes such as *Sarcocornietea fruticosae*.

In particular, even though little attention has been paid to the vegetation communities with medium values of biomass and with widespread distribution (Table 14.1), these ones deserve emphasis as they face desertification threats. This is the case of the salt-tolerant communities of *Pegano-Salsoletea*, *Crithmo-Staticetea*, and *Sarcocornietea fruticosae* along the coasts of Italy and its major islands.

Also summer-deciduous maquis communities referred to the alliance *Periplocion angustifoliae* are worth to be studied adequately, because they colonize the harshest areas of Italy (i.e. the coasts and southern Sicily and some of its satellite islets) and are prone to severe stress factors.

Another knowledge gap which should be filled concerns the shrubby communities colonizing the top of Mediterranean mountains; during recent decades, the role of these low (and slow) growing communities in carbon storage has increased along with the surfaces they occupy due to a strong reduction in traditional mowing and grazing practices, which has enhanced succession on a large scale.

If we take into account their distribution and their average biomass values, the deciduous shrubberies ascribed to *Prunetalia spinosae* and the 'true maquis' referred to *Quercetalia calliprini* play a prominent role among Italian non-forest woody vegetation.

At the moment it is still difficult to disentangle the effect of vegetation structure, dominat WS size and local bioclimate: for example, when calculated through dominant WS or by using the entire communities, their biomass values show substantial variation (Table 14.1). This is even true when the values concern the same study area but are derived from different papers; for instance, the values for the maquis biomass of Castelporziano range from 6,100 to 9,000 g/m^2 in Gratani et al. (1980) and from 600 to 3,000 g/m^2 in Gratani and Crescente (2000).

Thus, in order to avoid underestimating the biomass variability (cfr. *Erica arborea* in Table 14.2) of maquis and mantle communities with high structural (average height, dominant growth form, etc.) and floristic (species and WS richness) complexity, field measurements on the biomass of these communities should be made up to the association level. This topic is a stimulating challenge for future research and is necessary.

Unfortunately, many interesting papers such those of Merino and Martín Vicente (1981); Fioravanti (1999), Gratani and Crescente (2000) and Sternberg and Shoshany (2001), could not be used to create reliable models as they contained unsatisfactory community description or because they reported biomass data concerning several mixed vegetation units. In order to avoid such drawbacks, future investigations on Italian (and Mediterranean) shrubland plant communities should provide a more thorough description of both floristic and structural patterns leading to an univocal identification of phytosociological units (at least at class level).

Many papers (e.g. De Dato et al. 1999; Blanco Oyonarte and Navarro Cerrillo 2003) do not report raw data but provide algorythms or allometric equations which could be applied to forecast the biomass range of many Mediterranean woody species. In many other cases it is hard to build up models because in order to measure the trunk or stemwidth at its base the use an accurate gauge is needed, and this procedure becomes very difficult in some Mediterranean shrubs due to their irregular shape (Usò et al. 1997). As a matter of fact, the Ab of many Mediterranean shrubs shows a very wide range of variation (Table 14.2), as it is strongly influenced by progressive succession processes (Calvo 2007) and by the nature, the intensity and the frequency of disturbance factors (Castro and Freitas 2009). Thus, direct measuring of Ab and Bb of each vegetation unit and dominant WS should be encouraged and intensified.

Monitoring activities concerning biomass increase are recommended as well: in fact, reference data on this topic are so limited and variable that it is not possible to confidently estimate the annual growth of shrubland communities. However, the few available data on shrublands subject to progressive succession (e.g. Calvo 2007) suggest that if low or no disturbance occurs, many of these plant communities would rapidly become forest communities. Thus, high annual growth is expected.

Future investigations should reduce the present lack of data (Table 14.2) in order to improve our knowledge on the shrubland communities.

References

Bianchi L, Calamini G, Gregari E, Paci M, Pallanza S, Pierguidi A, Salbitano F, Tani A, Vedele S (2002) Valutazione degli effetti del rimboschimento in zone aride della Sardegna, risultati preliminari sulla vegetazione. It For Mont 4:353–368

Biondi E, Blasi C (eds) (2009) Manuale Italiano di Interpretazione degli Habitat della Direttiva 92/43 CEE (http://vnr.unipg.it/habitat/)

Blanco Oyonarte P, Navarro Cerrillo RM (2003) Aboveground phytomass models for major species in shrub ecosystems of Western Andalucia. Inv Agrar, Sistemas y Recursos Forestales 12(3):47–55

Calvo PP (2007) Phenology, biomass and community composition changes in European shrublands submitted to experimental warming and drought. PhD thesis, Universitat de Barcellona, Departament de Biologia Animal, Biologia Vegetal i Ecologia, Bellaterra, p 237

Castro H, Freitas H (2009) Above-ground biomass and productivity in the montado, from herbaceous to shrub dominated communities. J Arid Envir 73:506–511

Catarino FM, Correia OCA, Correia AI (1982) Structure and dynamics of Serra da Arrábida Mediterranean vegetation. Ecol Medit 8(1–2):203–222

Cherubini P, Gartner BL, Tognetti R, Bräker O, Schoch W, Innes JL (2003) Identification, measurement and interpretation of tree rings in woody species from Mediterranean climates. Biol Rev 78:119–148

Corona P, Calvani P, Mugnozza GS, Pompei E (2008) Modeling natural forest expansion on a landscape level by multinomial logistic regression. Plant Biosyst 142:509–517

Corona P, Ferrara A, La Marca O (1997) Sustainable management of forest for atmospheric CO_2 depletion, the Italian case. J Sustain For 5(3–4):81–91

Corona P, Pasta S, Giardina G, La Mantia T (2012) Assessing the biomass of typical Mediterranean shrub communities biomass of Mediterranean shrubs. Plant Biosyst 146(2):252–257

Costa G, La Mantia T (2005) Il ruolo della macchia mediterranea nello stoccaggio del carbonio atmosferico. Foresta@ 2(4):378–387. http://www.sisef.it/ (online)

Curcó A, Ibáñez C, Day JW, Prat N (2002) Net primary production and decomposition of salt marshes of the Ebre Delta (Catalonia, Spain). Estuaries 25(3):309–324

De Dato GD, Loperfido L, De Angelis P, Valentini R (1999) Establishment of a planted field with Mediterranean shrubs in Sardinia and its evaluation for climate mitigation and to combat desertification in semi-arid regions. iForest 2:77–84. http://www.sisef.it/iforest/show.php?id=498 (online 2009-06-10)

Fioravanti R (1999) Produttività ed allocazione della biomassa in una lecceta della Tenuta Presidenziale di Castelporziano. Tesi di Laurea, Università degli Studi della Tuscia Viterbo

García LV, Marañón T, Moreno A, Clemente L (1993) Above-ground biomass and species richness in a Mediterranean salt marsh. J Veg Sc 4:417–424

Gratani L, Amadori M, Veri L, Bruno F, Porri M (1980) Determinazione di un metodo di stima della biomassa nella macchia di Castelporziano (Lazio). Ann Bot 41(1):131–151

Gratani L, Crescente MF (2000) Map-making of plant biomass and leaf index for management of protected areas. Aliso 19(1):1–12

Hilbert DW, Canadell J (1995) Biomass partitioning and resource allocation of plants from Mediterranean-type ecosystems, possible responses to elevated atmospheric CO_2. In: Moreno JM, Oechel WC (eds) Global change and Mediterranen-type ecosystems, Ecological Studies 117. Springer, New York, pp 76–101

La Mantia T, Rühl J, Pasta S, Campisi D, Terrazzino G (2008) Structural analysis of woody species in Mediterranean old fields. Plant Biosyst 142(3):462–471

Laffoley (d'A) D, Grimsditch G (eds.) (2009) The management of natural coastal carbon sinks. IUCN, Gland, Switzerland, p 53

Margaris NS (1976) Structure and dynamics in a phryganic (East Mediterranean) ecosystem. J Biogeogr 3:249–259

Merino J, Martín Vicente A (1981) Biomass, productivity and succession in the scrub of the Doñana Biological Reserve in Southwest Spain. In: Margaris NS, Mooney HA (eds.) Components of productivity of Mediterranean-climate regions. Basic and applied aspects, pp 197–203

Montès N, Bertaudière-Montes V, Badri W, El-Zaoui H, Gauquelin T (2002) Biomass and nutrient content of a semi-arid mountain ecosystem, the *Juniperus thurifera* L. woodland of Azzaden Valley (Morocco). For Ecol Manage 166:35–43

Navarro Cerrillo RM, Blanco Oyonarte PB (2006) Estimation of above-ground biomass in shrubland ecosystems of southern Spain. Investigación Agraria Sistémas y Recursos Forestales 15:197–207

Neves JP, Ferreira LF, Simões MP, Gazarini LC (2007) Primary production and nutrient content in two salt marsh species, *Atriplex portulacoides* L. and *Limoniastrum monopetalum* L. in Southern Portugal. Estuaries and Coasts 30(3):459–468

Neves JP, Simões MP, Ferreira LF, Madeira M, Gazarini LC (2010) Comparison of biomass and nutrient dynamics between an invasive and a native species in a Mediterranean saltmarsh. Wetlands 30(4):817–826

Otto R, Fernández Palacios F, María J, Krüsi BO (2001) Variation in species composition and vegetation structure of succulent scrub on Tenerife in relation to environmental variation. J Veg Sci 12:237–248

Peressotti A, Asunis C, Cesaraccio C (1999) Stima della biomassa e della copertura vegetale di un ecosistema mediterraneo a macchia, il caso della riserva integrale Arca di Noè. Collana Tecnico-scientifica INAPA, Monitoraggio di un ecosistema a macchia mediterranea, misure ecofisiologiche e micrometeorologiche. Quaderno 6:81–86

Rapp M, Loissant P (1981) Some aspects of mineral cycling in the garrigue of southern France. In: Di Castri F, Goodall DW, Specht RL (eds) Mediterranean-type shrublands. Elsevier Scientific Publishing Company, Amsterdam, pp 289–301

Sağlam B, Küçük O, Bilgili E, Dinç Durmaz B, Baysal I (2008) Estimating fuel biomass of some shrub species (maquis) in Turkey. Turk J Agric For 32:349–356

Scarton F, Day JW, Rismondo A (2002) Primary production and decomposition of *Sarcocornia fruticosa* (L.) Scott and *Phragmites australis* Trin. ex Steudel in the Po Delta. Italy. Estuaries Coasts 25(3):325–336

Scarton F (2006) Produttività primaria epigea di sette alofite in laguna di Venezia. Boll Mus Civ St Nat Venezia 57(2):53–72

Sternberg M, Shoshany M (2001) Above-ground biomass allocation and water content relationship in Mediterranean trees and shrubs in two climatological regions in Israel. Plant Ecol 157:171–179

Usò JL, Mateu J, Karjalainen T, Salvador P (1997) Allometric regression equations to determine aerial biomasses of Mediterranean shrubs. Plant Ecol 132:59–69

The manufacturer's authorised representative in the EU is Springer Nature Customer Service Centre GmbH, Europaplatz 3, 69115 Heidelberg, Germany. If you have any concerns regarding our products, please contact ProductSafety@springernature.com

Printed and bound by CPI Group (UK) Ltd, Croydon, CR0 4YY
30/09/2026
02225602-0002